# Collins

**T0337526**

# ROAD ATLAS
# IRELAND
## TOURING EDITION
## Contents

Blue Flag & Green Coast beaches.......................................Inside front cover
Route planning map.........................................................................2-5
Places of interest.............................................................................6-12
Key to road map symbols & tourist information centres........................ 13
Road maps at 5.2 miles to 1 inch..................................................14-43
Administrative areas map and key to city & town plan symbols............. 44
City and town plans.......................................................................45-57
    Bangor...................................................................................45
    Belfast...............................................................................46-47
    Cork.......................................................................................48
    Drogheda ...............................................................................49
    Dublin ...............................................................................50-51
    Dún Laoghaire ........................................................................52
    Galway ...................................................................................53
    Killarney.................................................................................54
    Limerick.................................................................................55
    Londonderry (Derry) ...............................................................56
    Waterford ...............................................................................57
Index to place names .................................................................58-64
Distance chart & journey planning ................................. Inside back cover

Published by Collins
Westerhill Road, Bishopbriggs, Glasgow G64 2QT

www.harpercollins.co.uk

HarperCollins Publishers
Macken House, 39/40 Mayor Street Upper, Dublin 1, D01 C9W8, Ireland

Copyright © HarperCollins Publishers Ltd 2023
New edition 2023

Collins® is a registered trademark of HarperCollins Publishers Limited

Mapping generated from Collins Bartholomew digital databases

Ireland populations compiled from Census 2016. Source: Central Statistics Office
www.cso.ie Reproduced by permission.

Northern Ireland town populations derived from the 2011 Census. Source: Northern
Ireland Statistics and Research Agency www.nisra.gov.uk Reproduced by permission.

This product includes Intellectual Property from European National Mapping and Cadastral
Authorities and is licensed on behalf of these by EuroGeographics. Original product is
freely available at www.eurogeographics.org. Terms of the licence available at
http://www.eurogeograhics.org/form/topographic-data-eurogeographics.

Blue Flag beach information courtesy of The Blue Flag & An Taisce.
Green Coast beach information courtesy of Tidy Northern Ireland & An Taisce

The contents of this publication are believed correct at the time of printing. Nevertheless,
the publisher can accept no responsibility for errors or omissions, changes in the detail
given, or for any expense or loss thereby caused.

HarperCollins does not warrant that any website mentioned in this title will be provided
uninterrupted, that any website will be error free, that defects will be corrected, or that the
website or the server that makes it available are free of viruses or bugs. For full terms and
conditions please refer to the site terms provided on the website.

The representation of a road, track or footpath is no evidence of a right of way.

Printed in India     ISBN  978 0 00 859767 2    10 9 8 7 6 5 4 3

e-mail: roadcheck@harpercollins.co.uk

facebook.com/collinsref    @collins_ref

MIX
Paper | Supporting
responsible forestry
FSC™ C007454

This book contains FSC™ certified paper and other controlled
sources to ensure responsible forest management.

For more information visit: www.harpercollins.co.uk/green

## Key to map symbols

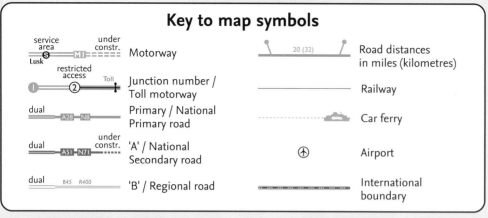

- Motorway
- Junction number / Toll motorway
- Primary / National Primary road
- 'A' / National Secondary road
- 'B' / Regional road
- Road distances in miles (kilometres)
- Railway
- Car ferry
- Airport
- International boundary

## Airport information

| Airport (airport code) & map ref | Dublin Airport (DUB) **29 E3** |
|---|---|
| Address | Fingal |
| | IRELAND |
| Republic of Ireland (UK) Tel Number | 01 (00353 1) 814 1111 |
| Web address | www.dublinairport.com |

**Belfast International Airport (BFS) 16 C4**
Aldergrove
Antrim
NORTHERN IRELAND
BT29 4AB
048 (028) 9448 4848
www.belfastairport.com

**City of Derry Airport (LDY) 15 F2**
Airport Road
Eglinton
Londonderry
NORTHERN IRELAND
BT47 3GY
048 (028) 7181 0784
www.cityofderryairport.com

**Connemara Airport (NNR) 25 E3**
Inverin, Galway
IRELAND
091 (00353 91) 593034
www.aerarannislands.ie

**Cork Airport (ORK) 42 B2**
Cork
IRELAND
021 (00353 21) 431 3131
www.corkairport.com

**Donegal Airport (CFN) 14 B2**
Carrickfinn
Kincasslagh
Letterkenny
Donegal
IRELAND
074 (00353 74) 954 8232
www.donegalairport.ie

**Dublin Airport (DUB) 29 E3**
Fingal
IRELAND
01 (00353 1) 814 1111
www.dublinairport.com

**George Best Belfast City Airport (BHD) 23 D1**
Belfast
NORTHERN IRELAND
BT3 9JH
048 (028) 9093 9093
www.belfastcityairport.com

**Ireland West Airport Knock (NOC) 26 B1**
Charlestown
Mayo
IRELAND
094 (00353 94) 936 8100
www.irelandwestairport.com

**Kerry Airport (KIR) 35 D4**
Farranfore
Killarney
Kerry
IRELAND
066 (00353 66) 976 4644
www.kerryairport.ie

**Shannon Airport (SNN) 35 F1**
Shannon
Clare
IRELAND
061 (00353 61) 712000
www.shannonairport.ie

**Waterford Airport (WAT) 38 C3**
Killowen
Waterford
IRELAND
051 (00353 51) 846600
www.waterfordairport.ie

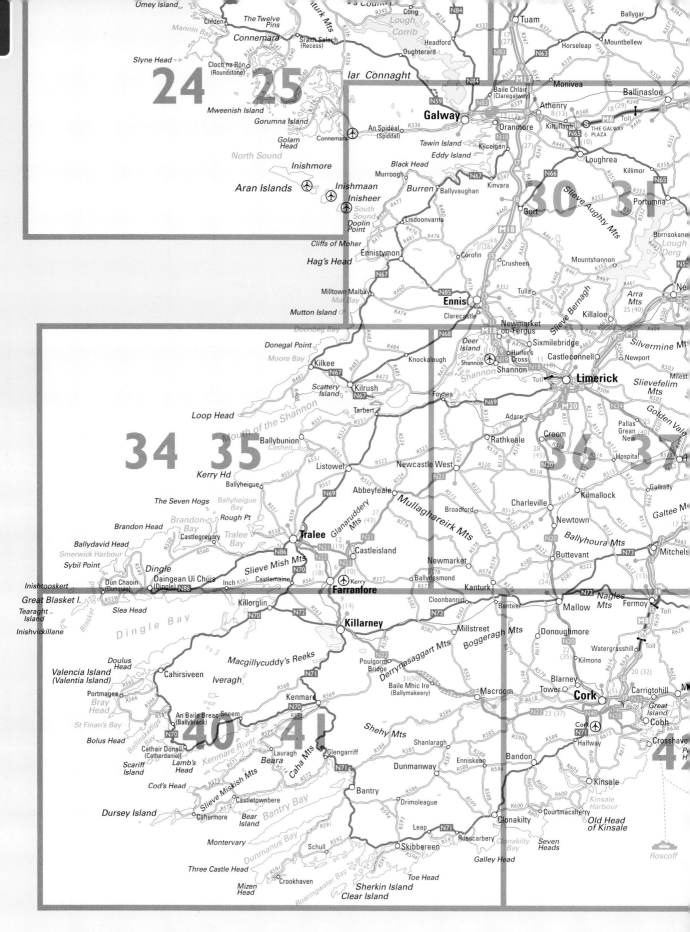

SCALE

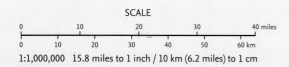

0      10      20      30      40 miles

0   10   20   30   40   50   60 km

1:1,000,000    15.8 miles to 1 inch / 10 km (6.2 miles) to 1 cm

M50 Dublin Port Tunnel: toll for non-HGV traffic only

# Ferry information

| Route & map ref | Belfast to Cairnryan 17 F3 |
| --- | --- |
| Duration | 2 hrs 15 min |
| Season | All year |
| Operator | Stena Line |
| Tel (Rep of Ire) | 01 907 5555 |
| Tel (UK) | 03447 707070 |
| Web address | www.stenaline.co.uk |

**Belfast to Cairnryan 17 F3**
2 hrs 15 mins
All year
Stena Line
01 907 5555
03447 707070
www.stenaline.co.uk

**Belfast to Douglas 17 F4**
2 hrs 45 mins - 3 hrs 15 mins
April - September
Isle of Man Steam Packet Co
0044 8722 992 992
08722 992 922
www.steam-packet.com

**Belfast to Liverpool
(Birkenhead) 17 F4**
8 hrs
All year
Stena Line
01 907 5555
03447 707070
www.stenaline.co.uk

**Buncrana to Rathmullan 15 E2**
40 mins
June - September
Lough Swilly Ferry
087 665 7522
00 353 087 665 7522
www.swillyferry.com

**Burtonport to Arranmore 14 B3**
15 - 20 mins
All year
Arranmore Ferry Company
07495 42233
00 353 7495 42233
www.arranmoreferry.com

**Carrigaloe to Glenbrook 42 B2
(Cork Harbour Ferry)**
5 mins
All year
Cross River Ferries
021 481 1485
00 353 21 481 485
crossriverferries.ie

**Cork to Roscoff 42 B2**
14 hrs
March - November
Brittany Ferries
021 427 7801
00 353 21 427 7801
www.brittanyferries.ie

**Dublin to Cherbourg 33 F1**
20 hrs
February - November
Irish Ferries
0818 300 400
00 353 818 300 400
www.irishferries.com

**Dublin to Douglas 33 F1**
2 hrs 55 mins
April - September
Isle of Man Steam Packet Co
0044 8722 992 992
08722 992 992
www.steam-packet.com

**Dublin to Holyhead 33 F1**
1 hr 50 mins - 3 hrs 15 mins
All year
Irish Ferries
0818 300 400
00 353 83 878 3862
www.irishferries.com

**Dublin to Holyhead 33 F1**
2 hrs 15 mins - 3 hrs 15 mins
All year
Stena Line
01 907 5555
03447 707070
www.stenaline.ie

**Dublin to Liverpool 33 F1**
8 hrs
All year
P&O Ferries
01 686 9467
01304 44 88 88
www.poferries.com

**Larne to Cairnryan 17 E3**
2 hrs 15 mins
All year
P&O Ferries
01 686 9467
01304 44 88 88
www.poferries.com

**Magilligan to Greencastle 16 A2**
15 mins
April - September
Lough Foyle Ferry Company
1800 938 004
00 353 83 878 3862
www.loughfoyleferry.com

**Passage East to Ballyhack 38 C3**
5 mins
All year
Passage East Car Ferry Company
051 382480
00 353 51 382480
www.passageferry.ie

**Rosslare to Cherbourg 39 E3**
18 hrs 30 mins
All year
Brittany Ferries
021 427 7801
0330 159 7000
www.brittany-ferries.co.uk

**Rosslare to Fishguard 39 E3**
3 hrs 15 mins
All year
Stena Line
01 907 5555
03447 707070
www.stenaline.co.uk

**Rosslare to Pembroke 39 E3**
4 hrs
All year
Irish Ferries
0818 300 400
00 353 818 300 400
www.irishferries.com

**Strangford to Portaferry 23 E2**
8 mins
All year
nidirect
00 44 300 200 7898
0300 200 7898
www.ndirect.gov.uk/strangford-ferry-timetable

**Tarbert to Killimer 35 E2**
20 mins
All year
Shannon Ferry Group Limited
065 905 3124
00 353 65 905 3124
www.shannonferries.com

## ⭐ ADARE VILLAGE & HERITAGE CENTRE    36 C2
### Limerick

The pretty village of Adare, which once belonged to the Kildare Fitzgeralds, and then the Earls of Dunraven, has several items of interest. The Gothic Revival manor house, now a luxury hotel, was built by the 2nd Earl in 1832 to his own design with additional work by James Pain and A.W. Pugin. The main hall is defined by stone arches and a handsome rococo staircase while the ornamental gardens are splendid for strolling. Medieval Desmond Castle, overlooking the River Maigue, still retains a fine square keep, as well as two great halls, a kitchen and stables. The most splendid remains in Adare are those of the 15th century Franciscan Friary (located in the grounds of the Adare Manor Golf Club) which can be admired from the long medieval bridge on the N20 to the north of the village. The story of Adare from the arrival of the Normans is told at the Adare Heritage Centre through audiovisual displays and models.

## ⛰ AILLWEE CAVE   Clare    30 B2

A short way south of Ballyvaughan, at the edge of the limestone plateau known as the Burren, Aillwee Cave, was discovered in 1940 by a local shepherd. Well over two million years old, the tunnel is filled with stalagmites and stalactites and an illuminated underground river and waterfall.

## ⭐ ARAN ISLANDS   Galway    25 D4

The islands of Inishmore, Inishmaan and Inisheer comprise the Aran Islands which lie at the entrance to Galway Bay. The main industries are tourism, sheep raising and fishing. Inishmore (*Inis Mór* meaning Big Island) is the largest, most populated and most frequently visited of the islands. The sandy beaches of Kilmurvey and Killeany contrast starkly with the sheer cliffs of the south of the island on which the ancient stone forts of Dún Aengus and Dún Eoghanachta sit. In Kilronan, the Aran Islands Heritage Centre (Ionad Árainn) introduces the archaeology, history, landscape and geology, culture and marine traditions of the islands. The islanders traditional dress is displayed and there are examples of the world famous Aran style knitting. The wreck of the Plassey on Inisheer (Inis Oirr) featured in the opening sequence of the 1990s sitcom 'Father Ted'.

The Plassey, Inisheer    Photo © Insuratelu Gabriela Gianina/Shutterstock

## 🏛 ARDMORE ROUND TOWER & CATHEDRAL    43 D2
### Waterford

Dominating the coastline, 12th century Ardmore Round Tower stands 29m (95ft) high on a monastic site which includes the ruins of a 12th century cathedral with a 9th century chancel. The west gable of the cathedral is decorated with a Romanesque frieze of carved figures depicting Biblical scenes and there are stones inscribed with Ogham writing, the first written Celtic language. Close by is a small stone church or oratory, possibly dating back to the 8th century, and reputed to be the site of St Declan's grave.

## 🏰 BALLINTOBER CASTLE   Roscommon    26 C2

Just outside the village of Ballintober is this ruin of a huge castle dating back to about 1300. It was the principal seat of the O'Connor dynasty from the time of the Anglo-Norman invasions until the 18th century when they moved to Clonalis. Many times besieged, rarely breached, it is without a keep, but retains its twin-towered gatehouse and corner towers.

## ✴ BATTLE OF THE BOYNE   Louth    29 E2

This famous battle took place on the 1st July 1690, when the deposed James II failed to regain the English throne from his son-in-law William of Orange at Oldgrange near Drogheda on the banks of the River Boyne. The site of the battle can be seen from a viewing point accessed by some steps opposite the Tullyallen turn off the N51. The visitor centre is on the south side of the river. Today, the battle is recalled on the 12th of July, as eleven days were lost when the change was made to the Gregorian calendar.

## ✝ BELFAST (ST ANNE'S) CATHEDRAL   Belfast City    46 B1

Close to the intersection of Clifton, Donegall and York Streets and Roy Avenue, St Anne's is the Protestant Cathedral for the dioceses of Connor an Down-and-Dromore. Built in 1898, in Romanesque style, it includes mosai by Gertrude Stein, has some very fine stained glass and is the burial place Lord Carson, the leader of the opposition to Home Rule who died in 1935.

## ⭐ BELLEEK POTTERY & VISITOR CENTRE    20 C1
### Fermanagh & Omagh

The pottery was established in the small village of Belleek in 1857 by Joh Caldwell Bloomfield who had inherited nearby Castle Caldwell. A keen amate potter, he noticed that all the necessary ingredients - feldspar, water, kaol and so on - were available locally. Within ten years, award winning high decorated, lustrous parian ware was being made and a tour of the attracti works demonstrates its continuing production. A shop, tearoom and museu are attached to the works.

## ✳ BIRR CASTLE GARDENS & SCIENCE CENTRE   Offaly    31 F2

Birr Castle has been occupied by the Parsons family since 1620 when S Laurence Parsons built most of the current building. Twice besieged in the la 17th century, the Gothic facade was added in the 19th century. Although th castle is only occasionally opened to the public, the gardens are open daily an consist of a beautifully landscaped collection of trees and shrubs, the tallest bo hedge in the world, and a kitchen garden, all set around a lake and waterfall An unusual feature is the case of the Great Telescope, built in the 1840s by th 3rd Earl of Rosse and the largest in the world until 1917.

## 🏰 BLARNEY CASTLE & BLARNEY STONE    42 B2
### Cork

The castle is just southwest of the village of Blarney. The keep, standing o a rocky outcrop, was built in 1446 by Cormac MacCarthy and then lost b the MacCarthys in the 17th century, finally passing to the Colthursts wh built the nearby 19th century house. The Blarney Stone lies just beneat the battlements. According to the rhyme 'A stone that whoever kisses, O h never misses to grow eloquent' but the origin of this bizarre piece of hokur is unknown, although it is said that Dermot MacCarthy was expert in usin honeyed language to keep the English at bay in the 16th century.

## ✝ BOYLE ABBEY   Roscommon    20 C4

The impressive remains of the 12th century Cistercian abbey known a Mainister na Buaille are on the north side of the market town of Boyle and ar the burial place of Ireland's most famous medieval religious poet, Donnchad Mor O'Daly, who may have been abbot here. Although badly damaged b Cromwell's army, the abbey is one of the best preserved in the country. Th church is a fine example of the transition from Romanesque to Gothic and ha particularly interesting carvings on the arcade capitals, whilst elsewhere th gatehouse, kitchen and cloister are all clearly in evidence.

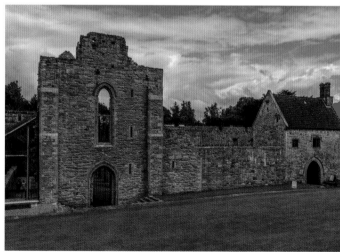

Boyle Abbey    Photo © jksz.photography/Shutterstoc

## ⭐ BRÚ NA BÓINNE (NEWGRANGE & KNOWTH TOMBS)    29 E2
### Meath

The Brú na Bóinne Visitor Centre is designed to present the archeologica heritage of the Boyne Valley, including the megalithic passage tombs o Newgrange, Knowth and Dowth, built some 3000 years BC. Newgrange is th largest and best known of the passage graves, being 80m (260ft) wide and u to 13m (43ft) high. A 24m (79ft) passage leads from the entrance to a cros shaped chamber with a corbelled roof of intricately carved slabs. The passag and chamber are illuminated by the light of the rising sun at Winter solstic (December 21st). Knowth is up to 15m (50ft) high and 85m (280ft) in diamete Surrounded by a number of other, smaller cairns, the main cairn contains tw tombs, the most westerly of which is 30m (100ft) in length. There is no direc access to the tombs. Visitors should use the minibus service that runs from Brú na Bóinne Visitor Centre. Visiting via formal tours only.

### BUNRATTY CASTLE   *Clare*                    30 C4

Bunratty Castle is one of Ireland's finest surviving tower houses. The magnificent keep has been restored to how it looked at the time of its construction on the banks of the Ratty in 1460. It was probably built by the MacNamaras but was soon in the hands of the O'Briens, who became the Earls of Thomond and who occupied the castle until 1712. Widely admired in its heyday, the three storeyed keep retains its corner towers and massive arches. Inside is the vaulted entrance hall, with so-called 'sheila-na-gig' (female fertility figure) in the wall, and chapel and cellars with 16th century stucco work. Many of the rooms are filled with a fine collection of period furniture, whilst mock banquets regularly evoke the castle's colourful past. At the base of the castle is the folk park, a collection of buildings and artefacts that illustrate Ireland's rural life at the turn of the century. Some of the buildings were moved here from an original site elsewhere whilst others are replicas. Traditional cottages from hill and valley areas of the Shannon region are represented, as well as a blacksmiths workshop, a mill, and an entire village as it would have been at the turn of the 19th century. Demonstrations of traditional skills are regularly given.

### BURREN CENTRE   *Clare*                    30 A3

The Burren (*boireann* meaning rocky place), is one of Ireland's most distinctive areas, being the largest area of Karst limestone in Europe. It almost resembles a lunarscape in its most isolated areas and is a stark contrast to the mountains or lush green landscape most associated with Ireland. The Burren Centre provides an introduction to the rich diversity of flora, fauna and geology of the Burren which has created this magnificent landscape.

The Burren                                Photo © Martin Fowler/ Shutterstock

### CAHIR CASTLE   *Tipperary*                    37 F3

A pretty town, Cahir boasts an impressive castle, the largest of its type in the country, which sits on a rocky outcrop in the middle of the River Suir at the foot of the Galty Mountains. It dates back to the 15th century and was inappropriately restored in 1840. However, behind its walls are a huge keep, a furnished great hall and two courts. Notwithstanding its solid appearance it was frequently overrun and in 1650 surrendered to Cromwell without coming to battle.

### CARRICK-A-REDE ROPE BRIDGE                    16 C1
*Causeway Coast & Glens*

This curiosity, north of Ballycastle, is a 20m (66ft) bridge of planks with wire rope handrails swinging 24m (80ft) above the sea and rocks separating Larrybane Cliffs from a small rocky island. There are magnificent views out to sea to be enjoyed during the intrepid journey across the bridge. This bridge has been erected here every spring for 200 years or more for the fishermen who operate a salmon fishery on the island.

### CARRICKFERGUS CASTLE   *Antrim & Newtownabbey*   17 E4

Carrickfergus dates back to the end of the 12th century but is very well preserved. Among the historical events associated with it are the landing of William of Orange in 1690 and the first action by an American ship in European waters in 1778. There remain three medieval courtyards within the walls containing a massive keep, with barrel-vaulted chambers and Great Hall. It contains a good selection of armour and armaments.

### CARROWKEEL MEGALITHIC CEMETERY                    20 B4
*Sligo*

About 6.5kms (4 miles) northwest of Ballinafad, just west of Lough Arrow, on the Bricklieve Mountains, is the remote site of Carrowkeel Bronze-Age passage-tomb cemetery. The setting consists of the site of an ancient village and 14 chambered cairns. It is not sure if the village, consisting of some 70 circular huts, and the burial ground, are from the same era but the cemetery, which contained cremated remains and was clearly planned, is undoubtedly one of the most important in the country.

### CASTLE COOLE   *Fermanagh & Omagh*                    21 E2

Restored by the National Trust, this elegant house, the seat of the Earls of Belmore, was designed in neo-Classical style by James Wyatt and Richard Johnston at the end of the 18th century. The facade is of Portland Stone whilst the interior plaster work, of classical simplicity, was undertaken by Joseph Rose in rooms still filled with their original furniture. Castle Coole is set in beautiful landscaped parkland with mature oak woodland and a lake which is home to a breeding colony of Greylag Geese.

Castle Coole                        Photo courtesy of The National Trust Northern Ireland

### CASTLE CALDWELL   *Fermanagh & Omagh*                    21 D2

The Caldwell family was responsible for the local porcelain industry, and their ruined 17th century castle is situated in a wooded peninsula on the banks of Lower Lough Erne. There are inspiring views across the lough from the gardens where wildfowl hides permit visitors to gaze undisturbed at the flocks of waterfowl that breed here. Waymarked walks and nature trails lead through beech and conifer forest to the shoreline.

### CASTLETOWN HOUSE   *Kildare*                    29 D4

Just outside the Liffeyside village of Celbridge, Castletown, built in 1722 for the Speaker of the Irish House of Commons, William Conolly, is perhaps the largest private house in Ireland. Consisting of a central block in early Georgian or Palladian style, flanked by colonnades, Castletown was designed by the Florentine, Alessandro Galilei, and Edward Pearce, architect of the Dublin Parliament. The house has been restored through the work of the Irish Georgian Society.

### CASTLE WARD   *Newry, Mourne & Down*                    23 E2

Situated in over 200ha (490 acres) of parkland, Castle Ward, now the property of the National Trust, is a half Gothic, half Palladian masterpiece built in 1765 for the 1st Viscount Bangor and his wife, each of whom favoured different styles. The family furniture is still in place, whilst the Trust has recreated a Victorian laundry. Close to the house is Old Castle Ward, a small Plantation castle built in 1610.

### CÉIDE FIELDS & VISITOR CENTRE   *Mayo*                    19 E1

Céide Fields consists of an area of 1500ha (3705 acres) containing field systems, enclosures and megalithic tombs dating back 5000 years. Preserved beneath the bogland of North Mayo and now coming to light through excavation, it is the most extensive Stone Age monument in the world. The visitor centre, an award winning pyramid structure of limestone and peat, has displays and exhibitions which interpret the geology, archaeology, botany and wildlife of the area and there are guided tours of the site.

Céide Fields Visitor Centre                        Photo © Mayo Naturally

### CHRIST CHURCH CATHEDRAL    *Dublin City*    50 B2

The cathedral church of the Protestant archdiocese of Dublin and Glendalough, and the principal religious and ceremonial church for the former English regime, is at the heart of medieval Dublin city. The first cathedral was built here in 1038 and a replacement begun over a century later. Various additions were made over the following centuries but after the nave vaulting collapsed in 1562 it went through a period of neglect. Eventually the whole church was remodelled in Gothic Revival style by G.E. Street. The interior has some fine surviving medieval sections including the groin-vaulted crypt and transepts, all of which date back to the 12th century, plus a sprinkling of interesting monuments including the casket containing the heart of St Laurence and a figure of Robert, 19th Earl of Kildare.

### CLIFFS OF MOHER & O'BRIEN'S TOWER    *Clare*    30 A3

The hustle and bustle of the car park are a marked contrast to the spectacular cliffs a short walk away which rise to over 200m (660ft) and extend for 8kms (5 miles). The 19th century O'Brien's Tower is a Victorian viewpoint on the highest point of the cliffs. The new visitor centre has been built into the hillside to blend in with the local contours and includes restaurant, exhibition centre, interactive displays and 'The Ledge' - a bird's eye CGI journey along the cliffs.

Cliffs of Moher      Photo © Stefano_Valeri/ Shutterstock

### CLONALIS HOUSE    *Roscommon*    26 C1

In the town that was the birthplace of Oscar Wilde's father, William Wilde, the noted antiquary, Clonalis House is the 19th century version of what was once the seat of O'Conor Don, of the O'Conors of Connacht, who produced two 12th century kings of Ireland. The house is a museum devoted to the family's colourful history.

### CLONMACNOISE MONASTERY    *Offaly*    27 E4

Sitting on a ridge on the banks of the Shannon, the remains of one of Ireland's first and most holy monasteries would ideally be approached by boat from Athlone. Founded in 545 by St Kiaran, a few months before his death, his tomb became an object of pilgrimage and the monastery grew to become a centre of Irish art and literature. The burial place of several Kings of Tara and Connacht, Clonmacnoise has endured many fires and numerous pillagings by Irish, Viking and English. The remains consist of two fine High Crosses, 400 memorial slabs from the 8th, 9th and 10th centuries, two Round Towers and eight churches. The 10th century West cross, with its frieze depicting St Kiaran and the local king from whom he obtained the monastery land, and the magnificently carved doorway of the Cathedral church are of particular interest.

Clonmacnoise Monastery      Photo © Pascal RATEAU/ Shutterstock

### CLOUGH CASTLE    *Newry, Mourne & Down*    23 E2

Clough is an Anglo Norman motte-and-bailey earthwork castle with added stone tower. It would originally have been defended by a timber palisade. Excellent views can be enjoyed from the top of the mound.

### COBH & THE QUEENSTOWN STORY    *Cork*    42 B2

Cobh (pronounced *Cove*) was renamed Queenstown after a visit by Queen Victoria in 1849 though it reverted to its original name in 1921. One of the world's largest natural harbours, it was a major departure point for emigrants especially during the famine years of the 1840s. On the promenade stands a monument to the victims of the Lusitania, torpedoed by a German submarine off the Kinsale coast in 1915. Cobh was also the last port of call for the Titanic before sinking in the Atlantic in 1912. The Cobh Heritage Centre in the town's restored Victorian railway station houses The Queenstown Story, a dramatic multimedia exhibition which covers Irish emigration, the transportation of convicts and the loss of the Lusitania.

### CONG ABBEY    *Mayo*    25 E2

Although the village is best known nowadays for the impressive mock medieval castle of Ashford, now a hotel, that was built for Sir Arthur Guinness, in Irish history Cong is associated with its monastery founded in the 6th century by St Feichin. This was superseded by an Augustinian monastery in the 12th century of which the inscribed base of a cross remains in the main street, whilst the remains of the chancel of the abbey church are to be found outside the village to the southwest. It has a beautiful north door but most appealing of all are the three doorways in the remaining convent buildings as these contain some of the finest medieval Irish carving in existence.

### CONNEMARA & CONNEMARA NATIONAL PARK
### VISITOR CENTRE    *Galway*    24 C2

West of Galway towards Clifden the landscape opens out into the majestic varied landscape of Connemara. The mountain ranges of the Maamturks and Twelve Pins (or Bens) rise out of the largely flat bog which sweeps down to the coastline, dotted with tiny lakes and islands and numerous golden beaches. The designated National Park area of this landscape comprises 2000ha (4945 acres) of scenic countryside including mountain, bog and grassland. The park is particularly rich in wildlife, there are herds of red deer, and the River Polladirk runs through the Glanmore valley at the centre of the park. There are several sites of archaeological interest. The visitor centre near Letterfrack has displays and an audiovisual presentation on the Connemara landscape. A signposted nature trail and three walking trails start at the centre and include views of Ballynakill Bay, Inishbofin and Inishark.

Connemara      Photo © MNStudio/ Shutterstock

### CRAG CAVE    *Kerry*    35 E4

Not far from Tralee, Crag Cave is really a network of limestone caves some 4kms (2.5 miles) in length, discovered only in 1983. Over a million years old, they are festooned with stalactites and stalagmites, many of which have conjoined to form curtains and pillars. The Crystal Gallery is so called because of the white straw stalactites that glitter in the light.

### CROAGH PATRICK & NATIONAL FAMINE MONUMENT
*Mayo*    19 D4

Known locally as the Reek, Croagh Patrick rises to 762m (2515ft). St Patrick is said to have spent 40 days on the summit in 441AD, fasting and praying for the people of Ireland. Since early Christian times a pilgrimage has taken place every July when the devout climb the mountain in the footsteps of the saint. A Celtic hill fort encircles the summit and a dry stone oratory is one of the oldest stone churches in Ireland. At the foot of Croagh Patrick stands a bronze monument to the Great Famine of the 1840s sculpted by John Behan depicting a 'coffin ship' with skeleton bodies in the rigging.

National Famine Monument

Photo © chrisdorney/ Shutterstock

### CRUMLIN ROAD GAOL  *Belfast*  23 D1
With daily guided tours, you can explore this 19th century prison. Designed by Charles Lanyon, it was constructed between 1843 and 1845. During its 150-year history, 17 men were executed here, the last in 1961. It operated as a prison until 1996, when it finally closed. It became known as the Alcatraz of Europe, although there were a handful of successful escapes from the prison over the years.

### DONEGAL CASTLE  *Donegal*  20 C1
The late 15th century castle is just off the main square of this market town that has given its name to the county. Originally the seat of the O'Donnells, the castle tower was built by Red Hugh II O'Donnell in 1505, whilst additions, including the magnificent fireplace in the banqueting hall, were made in the 17th century by a later occupant, the Planter Sir Basil Brooke, who was responsible for the fine Jacobean fortified house that forms part of the castle.

### DRUMCLIFFE HIGH CROSS & ROUND TOWER  *Sligo*  20 B2
The remains of the round tower and a 10th century high cross are all that is left of a monastic settlement founded by St Colmcille (St Columba) in 574. The monastery was flourishing in the 13th century but went into decline soon afterwards. Irish poet W.B.Yeats, who spent much of his childhood in Sligo, is buried in the churchyard here.

### DUBLIN CASTLE  *Dublin City*  50 B2
Built at the turn of the 13th century, Dublin Castle has been a gaol for many important characters throughout Irish history, as well as the seat of English power until 1922. Nonetheless, it has never had to withstand a serious assault. Its only real enemy was neglect and the whole edifice was more or less reconstructed throughout the 18th century. Although it still retains an air of fortified unity, the only substantial medieval remains are the southeast or Record tower and the layout of the Upper Castle Yard. The original moat has also been discovered. The 18th century work was the last major brick construction in Dublin and guided tours will take you through the magnificently decorated and furnished State Apartments which include St Patrick's Hall (where Irish presidents are inaugurated), the circular Supper Room and the Throne Room. In the Lower Castle Yard is the impressive Church of the Most Holy Trinity, or the Chapel Royal, in Gothick style.

### DÚN AONGHASA (DÚN AENGUS FORT)  25 D4
*Inishmore, Aran Islands, Galway*
Towards the far southwestern part of the island of Inishmore, on a spectacularly desolate slope overlooking the Atlantic, are the scattered remains of Dún Aonghasa (or, according to the most correct pronunciation, Doon Eeneece), one of the most important prehistoric forts in Europe, some 2500 years old. The 4.5ha (11 acre) site consists of three enclosures with dry-stone walls up to 5m (18ft) high. Most striking is the arrangement of jagged limestone menhirs which defend the middle wall.

### DUNLUCE CASTLE (or Mermaid's Castle)  16 B1
*Causeway Coast & Glens*
Just northwest of the town of Bushmills, famous for its whiskey distillery, Dunluce Castle is dramatically situated on a high rock overlooking the sea. It was first a Macquillan stronghold, then came to the MacDonnells until the 17th century when the occupied domestic quarters fell into the sea. The earliest part dates back to the late 13th century whilst the other sections - the Scottish gatehouse, the loggia - were built at different times right up to the 17th century.

### DUNSOGHLY CASTLE  *Fingal*  29 E3
Just 5km (3 miles) to the northwest of Finglas, Dunsoghly Castle consists of a stately residential keep-like tower on three floors buttressed by rectangular corner turrets. It was built in the 15th century by Thomas Plunkett, Chief Justice of the King's Bench, whilst the nearby chapel was built in 1573 by Sir John Plunkett, also Chief Justice but for the Queen's Bench.

### FOTA HOUSE, ARBORETUM & GARDENS  *Cork*  42 B2
Fota House is on Fota Island which, accessible by causeways, sits in the River Lee Estuary. Built in 1820 for the Earls of Barrymore, it is a handsome Regency house with a fine collection of 18th and 19th century furniture and Irish landscape paintings. In the surrounding grounds is the internationally renowned arboretum begun in the 1820s with its collections of rare shrubs and semi-tropical and coniferous trees. There is an orangery, a range of terraces and four walled gardens which include stone ornamental gateways and a summer house in the Italian Garden. Also in the grounds is a wildlife park (see below).

### FOTA WILDLIFE PARK  *Cork*  42 B2
Located in the grounds of Fota House but a separate attraction, Fota Wildlife Park holds over 90 species of animal including giraffes, bison and cheetahs and is important in the conservation and breeding of endangered species. Wherever possible the animals are encouraged to thrive in unrestricted areas and some, such as lemurs, kangaroos and wallabies, are allowed to roam free through the park.

### GALLARUS ORATORY  *Kerry*  34 A4
On the beautiful Dingle peninsula, at Lateevmore, which is about 3km (2 miles) east of Ballyferriter, is Gallarus oratory, a corbel-roofed dry-stone structure of remarkable perfection and completely waterproof. A Christian chapel that is over one thousand years old, it lacks only the crosses that once decorated the roof.

### GIANT'S CAUSEWAY  *Causeway Coast & Glens*  16 B1
This monumental terrace of steps stretches for almost 1km out into the ocean on the north Antrim coast between Port Ganny and Port Noffer. It resulted from gigantic outpourings of volcanic basalt some 60 million years ago. The rock cooled into two distinct formations. Firstly, a lower layer of thousands of regular hexagonal colums and secondly, an upper layer of slim uneven prisms. This amazing piece of coastline now belongs to The National Trust, who have created a coastal path from the Causeway to Dunseverick. The Visitor Centre has an exhibition and audiovisual theatre, which outlines the geological history of the Causeway.

### GLENDALOUGH  *Wicklow*  33 E3
Glendalough is a 6th century Christian monastic site founded by St Kevin. There are some interesting remains, namely the impressive 31m (100ft) high round tower, several stone churches and decorated crosses. The Visitor Centre is set at the gateway to Glendalough Valley and Wicklow Mountains National Park. It contains a model of the site as it would have appeared in 1080 and an audiovisual presentation.

### GRIANAN AILIGH  *Donegal*  15 E3
8km (5 miles) northeast of Newtowncunningham, on Greenan Mountain, this fort consists of a huge cashel (stone fort), probably dating from the early Christian period, standing at the centre of a series of three earthen banks which are either late Bronze or early Iron Age. An important stronghold of the Christian kingdom of Ailigh, it retained a mythological importance long after its strategic value had passed away. There are, however, still wonderful views across the Foyle and Swilly.

Giant's Causeway

Photo © Joe Gough  Used under license from Shutterstock.com

### HILL OF TARA  *Meath*                                    29 D3

In the late stone age, a passage tomb was built here and later, an iron age hill fort. It is best known, however, for being the seat of the legendary 'Kings of Ireland', reaching the height of its success with Cormac MacAirt in the third century. Its importance gradually declined with the coming of Christianity, although it was not finally abandoned until the 11th century. Excavations continue and there are guided tours and a visitor centre.

### HOOK LIGHTHOUSE  *Wexford*                              38 C4

This 13th century lighthouse is one of the oldest working examples in the world and was opened to the public when it was fully automated in 1996. 115 steps snake up the two tiered 36.6m (120ft) high building, almost all of which is original. The cafe and craft shop are situated in the former keeper's house.

### IRISH NATIONAL STUD, JAPANESE GARDENS & ST FIACHRA'S GARDEN  *Kildare*                              32 C2

The 400ha (1000 acre) farm at Tully has been used as a stud farm since the turn of the century. Guided tours run throughout the day and there is a museum and video showing the birth of a foal. In the grounds of the National Stud, the Japanese Gardens were planted by the Japanese gardener, Eida, and his son Minoru, between 1906 and 1910. Symbolising the ages of man from birth to death, the route takes you from the Cave of Birth via the Hill of Ambition and the Well of Wisdom to the Gate of Oblivion. The Zen Meditation Garden was added in 1976 and the Garden of Eternity was added in 1980. St Fiachra's Garden was created as a millennium project to celebrate the Irish landscape in its most natural state and contains woods, wetlands, lakes and islands.

### JERPOINT ABBEY  *Kilkenny*                              38 B2

The remains of Jerpoint Abbey are a superb example of Cistercian monastic life, in that many of the domestic areas are still recognisable. Founded in the late 12th century, it had its own gardens, kitchens, infirmary, watermills, granary and cemetery. The Irish-Romanesque transepts and chancel are the oldest parts, where faded wall paintings can still be seen. The central tower is a 15th century addition. The cloister piers have been restored and show carvings representative of drawings found in Medieval manuscripts.

St Fiachra's Garden                          Photo © Irish National Stud

### KELLS PRIORY  *Kilkenny*                                38 B1

In 1193 a priory was founded in Kells for Canons Regular of St Augustine from Bodmin in Cornwall. The impressive remains, 2ha (5 acres) surrounded by substantial medieval fortified walls, with mostly complete 15th century dwelling towers, are divided into two courts by a branch of the river, in the northernmost of which are the remains of the church, with traces of medieval paving tiles, and the ruined claustral buildings.

### KILLARNEY NATIONAL PARK  *Kerry*                        41 D1

The National Park covers an area of over 10,000ha (24,700 acres) most of which is given over to the beautiful three lakes of Killarney. There are extensive areas of natural oak and yew woodland and the park contains the only remaining native red deer population in Ireland. The Visitor Centre at Killarney House gives information on the flora and wildlife in the Park, including Killarney Fern, Strawberry Tree, Northern Emerald Dragonfly and a flock of Greenland Whitefronted Geese which reside in the bogs over the winter months. There is also an information point at Muckross House a short walk from the 18m (60ft) high Torc Waterfall. Ladies View is a viewpoint on the N71 near Upper Lake so named because of the pleasure it gave to Queen Victoria's ladies-in-waiting in 1861.

Killarney National Park          Photo © Paul Merrett  Used under license from Shutterstock.com

### KILMAINHAM GAOL  *Dublin City*                          29 E4

The imposing building at Kilmainham dates from the 1790s and spent 130 years as a prison. After its closure in 1924 it re-opened in 1966 as a museum dedicated to the Irish patriots imprisoned there from 1792 – 1924. Patrick Pearse and James Connolly were executed in the prison yard and Eamon de Valera, later Prime Minister and then President of Ireland, was one of the last inmates. It is one of the largest unoccupied gaols in Europe with tiers of cells and overhead catwalks. Access is by guided tour which includes the punishment cells, hanging room and prison yard, and there are regular temporary exhibitions and special events.

### KING JOHN'S CASTLE  *Louth*                             22 C4

The small seaport of Carlingford, on the unspoilt Cooley Peninsula, is located at the foot of the 590m (1935ft) Slieve Foye, overlooking Carlingford Lough. The castle remains, strategically located to command the quay, date back to the late 12th century and played host to King John who stayed here on his way to attack Hugh de Lacy, at Carrickfergus. It has an unusual D-shape while the west gateway was designed to allow the entry of only one horseman at a time. The remains of an earlier castle include the southwest tower and the west wall.

### KING JOHN'S CASTLE  *Limerick*                          55 B1

In the Old Town of Limerick (on a sort of island formed by the River Shannon and the River Abbey), the 13th century castle, the most formidable English stronghold in western Ireland, is a fine example of Norman fortified architecture. It has recently been partly converted into a museum which displays antique armaments (catapults and battering rams) and tells of the castle's role in Limerick's dramatic history. There is also an interactive exhibition.

### KNOCK BASILICA & SHRINE  *Mayo*                         19 F4

One of the great Marian shrines of the world, Our Lady's Shrine at Knock attracts over 1.5 million visitors each year. An apparition of the Virgin Mary, St Joseph and St John on 21 August 1879 was witnessed by 15 local people on the south gable of the parish church of St John the Baptist. Ever since, Knock has been a place of pilgrimage. The focal point of the shrine is the gable of the apparition and the shrine Oratory. Nearby is the Basilica of Our Lady built in 1976 to accommodate the flow of pilgrims. It is the largest church in Ireland with room for up to 20,000 people. Pope John Paul II visited Knock on his tour of Ireland in September 1979.

### LISMORE CASTLE GARDENS  *Waterford*                     43 D1

The castle, handsomely located above the Blackwater, was built in the 19th century by Joseph Paxton, architect of the Crystal Palace in London, for the 6th Duke of Devonshire, incorporating the remains of the medieval castle erected by Prince John of England in 1185. The castle (private) is partially surrounded by delightful walled gardens, with areas of woodland, shrubberies, and a Yew Walk. In spring the gardens are at their best when the camellias and magnolias are in flower. The Elizabethan poet Edmund Spenser is said to have composed part of the Faerie Queene in the grounds.

### LISSADELL HOUSE  *Sligo*                                20 B2

Just over 6km (4 miles) outside the hamlet of Raghly which hangs off a promontory on the north side of Sligo Bay, Lissadell House, situated in conifer clad parkland overlooking the sea, was built in 19th century Classical style for the patriotic Gore-Booth family. The Arctic explorer Sir Henry Gore-Booth was born here, as were his daughters Eva, the poetess, and Constance, who became the first woman member of the British House of Commons, but who chose to sit instead in the revolutionary Dail Eireann as minister for Labour. Refreshments are still served in the old-fashioned kitchen.

### MARBLE ARCH CAVES  *Fermanagh & Omagh*  **21 D3**

This system of caves, about 16km (10 miles) southwest of Enniskillen, has been formed by the action of three streams on a bed of Dartry limestone on 667m (2188ft) Mount Cuilcagh. Underground they converge to form a single river, the Cladagh, which flows via the 9m (30ft) limestone Marble Arch into Lough Macnean. The tour of the cave includes a boat ride on the underground lake and the presentation of an array of imaginatively illuminated and named rock formations, one of which, a stalactite, is over 2m (7ft) long.

### MELLIFONT ABBEY  *Louth*  **29 D2**

6km (4 miles) northwest of Drogheda, Mellifont was Ireland's first Cistercian abbey. Founded in 1140 by the King of Uriel, at the instigation of St Malachy, who had been inspired by St Bernard's work at Clairvaux, the abbey became the home of the Moore family after its suppression in 1539. Though scattered, the remains are of great interest and include portions of the original Romanesque cloister arcade, the 13th century chapter house extension, and the octagonal washroom.

### MONEA CASTLE  *Fermanagh & Omagh*  **21 D2**

About 9.5km (6 miles) northwest of Enniskillen, Monea Castle was built in 1618 by the Rev. Malcolm Hamilton in Scottish Plantation style. It was burnt out in the 18th century but is in a reasonable state of preservation and its remaining two circular towers at the front and its crow-stepped gables add to its Scottish flavour.

### MOUNT STEWART HOUSE  *Ards & North Down*  **23 E1**

The 18th century former seat of the Marquess of Londonderry, and childhood home of Lord Castlereagh, the 19th century British Prime Minister, now belongs to the National Trust. A severely Classical building, it sits amid 32ha (80 acres) of gardens renowned for their many rare plants, trees and fanciful topiary. The Temple of the Winds, also in the grounds, is an exquisite banqueting hall built in 1785.

Mount Stewart House                    Photo courtesy of National Trust Northern Ireland

### MOUNT USHER GARDENS  *Wicklow*  **33 F3**

Next to the village of Ashford, charmingly located by the River Vartry, the gardens of Mount Usher are made up of 8ha (20 acres) planted with over 5000 species of flora. Including many sub-tropical plants, the naturalised gardens, laid out in 1868 by Edward Walpole, of a Dublin family of linen manufacturers, are famous for the Eucalyptus and Eucryphia collections and offer some fine woodland walks.

### MUCKROSS HOUSE  *Kerry*  **41 D1**

A Victorian mansion built in 1843 in the Elizabethan style. The interior shows not only the lifestyle of the gentry of the period but also that of the servants. Grounds have water gardens, colourful displays of azaleas and rhododendrons in the spring and a limestone rock garden. There are also craft workers demonstrating weaving, bookbinding and pottery and the Killarney National Park Visitor Centre. Three working farms in the grounds use the traditional farming methods of the early 20th century.

### MURLOUGH NATIONAL NATURE RESERVE  **23 D3**

*Newry, Mourne & Down*

Murlough is an area of sand dunes (some of which were formed over 5000 years ago), heath, grassland and woodland surrounding the estuary of the Carrigs River and the shore of Dundrum Bay. With the diversity of habitats there is a wide range of plants, birds and wildlife including badgers and stoats. The estuary attracts many species of wader, duck and geese, migration times being of particular interest. Common and grey seals are visitors to the beach. A National Trust property.

Muckross House                    Photo © Bildagentur Zoonar GmbH/Shutterstock

### NATIONAL MUSEUM OF COUNTRY LIFE  *Mayo*  **19 F4**

The only branch of the National Museum of Ireland to be situated outside Dublin and housed in modern purpose-built exhibition galleries, the collection portrays the social history of rural Ireland with the primary focus between 1850 and 1950. The lives of the Irish people are presented in the context of history, folklore and the natural environment. The new galleries are adjacent to the restored Victorian Gothic style house of Turlough Park and overlook the lake and formal gardens.

### NATIONAL MUSEUM OF IRELAND
### (MUSEUMS OF NATURAL HISTORY, ARCHAEOLOGY
### & HISTORY AND DECORATIVE ARTS & HISTORY)

*Dublin City*  **50 A2 / 50 C2**

Ireland's premier cultural institution has a strong emphasis on the country's art, culture and natural history. There are collections and exhibitions at three sites in Dublin – natural history at Merrion Street, national antiquities in Kildare Street, and decorative arts, economic, social and military history at Collins Barracks, Benburb Street.

### NATIONAL BOTANIC GARDENS  **29 E4**

*Dublin City*

Dublin's Botanic Gardens, in the suburb of Glasnevin, beyond the Royal Canal, were created in 1795 by the Royal Dublin Society, passing to the state in 1878. In its 20ha (49 acres) there are several notable conservatories including one, over 122m (400ft) in length, by Richard Turner who also designed for Kew in London. One of the finest botanical gardens in Europe, it is especially noted for its conifers, cycads, herbaceous borders and orchids, and boasts a lily pond, a sunken garden and a rock garden.

### NAVAN FORT  *Armagh City, Banbridge & Craigavon*  **22 B2**

A little way to the west of the town of Armagh, the remains at Navan are of an 7ha (18 acre) hill fort, crowned by a ceremonial tumulus, which together form the last remains of the seat of Ulster kings between 350 BC and 332 AD. It is also the Emhain Macha, the legendary home of Cu Chulainn, one of the knights of Ulster mythology.

### NESS WOOD COUNTRY PARK  *Derry City & Strabane*  **15 F3**

The 19ha (47 acres) of woodland were originally dominated by oak trees but many other species were added from the 17th century. The highlight of Ness, however, is the spectacular 9m (30ft) waterfall, part of the River Burntollet, which since the last Ice Age has also created a series of gorges and rapids through the metamorphic rock.

### PARKE'S CASTLE  *Leitrim*  **20 B3**

On the banks of Lough Gill, in Kilmore, close to Dromahaire, Parkes Castle is a Plantation castle in markedly Scottish style built in the 17th century on the site of an earlier castle. In a good state of preservation, its courtyard walls are fortified with picturesque towers and gatehouses and it played a key role in the war of 1641 - 52. There is a permanent exhibition about the area as well as an excellent audiovisual show.

### POWERSCOURT HOUSE & GARDENS  *Wicklow*  **33 F2**

Powerscourt House was built in 1730 for Viscount Powerscourt by the Huguenot architect Richard Cassels and then enlarged and altered in the 19th century. Its mountain setting is magnificent, as are the gardens with their handsome 19th century terraces, Monkey Puzzle Avenue, and Japanese Garden, added in 1908. In the grounds, and approachable by a separate car entrance, is a spectacular 120m (400ft) waterfall.

Powerscourt Garden    Photo © Wil Tilroe-Otte  Used under license from Shutterstock.com

### PROLEEK DOLMEN  *Louth*                                  22 C4
To the north of Dundalk, the capital of Louth, in Aghnaskeagh, are two prehistoric cairns and a fort. Nearby Proleek is the site of the so called 'Giant's Load', a tomb that is the legendary grave of Para Bui Mor Mhac Seoidin, the Scottish giant who challenged Finn MacCool. A trio of smaller upright stones supporting a larger capstone, the tomb dates back to 3000 bc. It is thought that the capstone was hauled into position by means of an earthen ramp that has now vanished.

### RING OF KERRY  *Kerry*                                   40 B1
The Ring of Kerry is a famous circular scenic route of about 185km (115 miles) around the Iveragh Peninsula. Clearly it can begin at any point on the route but the town of Killarney is generally considered the gateway to the peninsula, even if the best section lies between Kenmare and Killorglin. The fine mountain and maritime scenery is a constant companion but the route passes through or near to a number of interesting places including Sneem, the 2000 year old fort at Staigue; Derrynane National Historic Park, the resort of Waterville; and Glenbeigh with its Bog Village Museum. Whilst the principal route more or less follows the coast, some of the finest scenery is to be found along the unmarked roads running through the interior of the peninsula.

### ROCK OF CASHEL  *Tipperary*                              37 F4
One of the most spectacular sights in Ireland, the Rock of Cashel is a steep limestone outcrop surmounted by the ruins of the ancient capital of the Kings of Munster. According to legend, St Patrick baptised Corc the Third here; and Brian Boru, High King of Ireland, was crowned here in 977. In 1101 King Murtagh O'Brien donated the rock to the church after which it became the See of the Archbishopric of Munster. The ruins are extensive and fascinating. Cormac's Chapel, built in the 1130s in a style sometimes called Hiberno-Romanesque, contains a magnificent carved 11th century sarcophagus; whilst the carved Cross of St Patrick is set into the coronation stone of the Kings of Munster. The main cathedral is essentially 13th century and although it has suffered pillage and neglect, it remains a fine example of Irish Gothic. There are remarkable views across the surrounding countryside from the part of the cathedral known as the Castle, built to house the 15th century bishops.

### ROCK OF DUNAMASE  *Laois*                                32 B3
Close to Portlaoise, on the Stradbally road, a large ruined castle sits on top of this 46m (150ft) rock, all that remains of what was a considerable fortress, destroyed in the Cromwellian wars. Through the marriage of the daughter of the King of Leinster, it had moved into Anglo Norman hands in the late 12th century, and was twice rebuilt, in 1250 and at the end of the 15th century. There are fine views to be had from the summit of the rock where the remains of the gatehouse, walls and 13th century keep are still in evidence.

### RUSSBOROUGH HOUSE  *Wicklow*                             33 D2
One of Ireland's foremost Palladian houses was built for a Dublin brewer, Joseph Leeson, 1st Earl of Milltown, by Richard Cassels and Francis Bindon in the 1740s. A granite exterior conceals an interior coated in extravagant stucco work and bearing a fine painting collection.

### ST CANICE'S CATHEDRAL  *Kilkenny*                        38 B1
The city of Kilkenny actually takes its name from St Canice who founded a monastery here in the 6th century, upon the site of which stands the current cathedral. Much restored in the 19th century, it is the second largest medieval cathedral in Ireland. Inside is the finest display of burial monuments in the country whilst the high round tower adjacent to the cathedral is the only substantial relic from the monastery. The building is essentially 13th century and the oldest tomb is also from this period, although the oldest decipherable slab is that of Jose Kyteler, the father of Alice Kyteler, who was tried for witchcraft in 1323.

### ST PATRICK'S CATHEDRAL  *Dublin City*                    50 B3
St Patrick's Cathedral is the largest church in Ireland and has for much of its existence had to compete with Christ Church for its Cathedral status. Founded in 1191 and rebuilt in the 14th century it is essentially Early English in style, although damage wrought during the Cromwellian wars was only restored in the 1860s. The satirist Jonathan Swift, Dean of the Cathedral 1713–45, is buried here. The baptistry, paved with 13th century tiles, is all that remains of the first church, whilst at the west end of the nave stands the old Chapter

House door. The best preserved part of the medieval church, however, i[s] Choir which came to be the chapel of the now defunct Order of St Pa[trick] instituted in 1783. Among the monuments perhaps the most impressive i[s] Boyle Monument, erected by the Earl of Cork in memory of his wife Cathe[rine]

### SCRABO COUNTRY PARK  *Ards & North Down*                 23
The local landscape is dominated by Scrabo Hill, a layer of volcanic rock [over] a mound of sandstone dominates the landscape. It is crowned by Sc[rabo] Tower, erected in 1857 in memory of the third Marquess of Londond[erry] Built in saturnine dolerite, and sandstone, it now houses a museum abou[t the] surrounding countryside, some of which is made up of 19th century b[roadleaf] and mixed woodland. There are magnificent views across Strangford Lou[gh]

### SEA LIFE BRAY AQUARIUM  *Wicklow*                        33
Features marine life from the seas around Ireland including stingrays, co[nger] eels, and sharks; also freshwater fish from Irish rivers and streams. A to[uch] pool (check if open) gives children the opportunity to pick up small creat[ures] such as starfish, crabs and sea anemones. By way of contrast is the fascina[ting] 'Toxic Terrors' zone with many sea creatures from around the world w[hich] have proved harmful or fatal to humans.

### STAIGUE STONE FORT  *Kerry*                               40
Amid the beautiful scenery overlooking Kenmare Bay, Staigue is a stone 2[000] year old ringfort made up of a 4m (13ft) thick rampart divided into terr[aces] and linked by a system of stairways.

### STROKESTOWN PARK - THE IRISH NATIONAL FAMINE MUSEUM
*Roscommon*                                                  27
Designed by Richard Cassells for the Mahon family, Strokestown Park Ho[use] is a restored 18th century neo Palladian mansion and most of its orig[inal] furnishings and family possessions have been retained. The grounds inc[lude] the longest herbaceous border in Ireland, rose garden, large ornamenta[l] pond, Georgian peach house and vinery and a 17th century tower. The Fam[ine] Museum is housed in the former stable yards and has an extensive collec[tion] of estate documents including letters and pleas written by the tenants. M[ajor] Dennis Mahon, landlord of Strokestown during that period, was assassin[ated] after he had tried to clear two thirds of his destitute tenants from the esta[te]

### THE GOBBINS CLIFF PATH  *Mid & East Antrim*              31
The Gobbins Cliff Path follows a spectacular route which hugs the cliff f[ace] The fully guided walk takes around 2.5 hours, taking in breathtaking sce[nery] and local wildlife. It is described as an arduous trek, with narrow and une[ven] pathways requiring walking boots.

### THOOR BALLYLEE  *Galway*                                  30
The 16th century tower of Ballylee Castle, the Thoor Ballylee of Yeat's poe[m] is 8km (5 miles) northeast of the market town of Gort. A charming ivy [clad] tower on the banks of the river, Ballylee was Yeats' home in the 1920s, wh[ere] he wrote the volume of poems entitled The Tower. After he left in 1929 [the] tower fell into disrepair once more until its restoration as a Yeats Muse[um] in 1965.

### TITANIC BELFAST  *Belfast City*                           23
An iconic building in the heart of Belfast's docklands, Titanic Belfast prov[ides] visitors with a unique insight into all things Titanic. It tells the story of Tit[anic] from her construction in Belfast to her fateful maiden voyage, using stunn[ing] special effects, innovative interactive features and full-scale reconstruction[s]

### WESTPORT HOUSE, GARDENS & PIRATE ADVENTURE PARK
*Mayo*                                                       19
Adjacent to the pretty village of Westport, planned by James Wyatt for [the] Marquess of Sligo, the house, built in about 1730, is by Richard Cassels, w[ith] additions by Wyatt in 1778. The house, entered from the Quay, contain[s a] mixture of antique silver and furniture and modern entertainment facili[ties] whilst in the demesne itself is an ornamental lake, created by controlling [the] tides of Clew Bay, and a miniature zoo.

### WICKLOW MOUNTAINS NATIONAL PARK                          33
*Wicklow*
Established in 1991, the total managed area is now 20,000ha (49,400 ac[res]) with a core area based around the lakes at Glendalough. The Park inclu[des] Liffey Head Bog and the Glenealo Valley Heath and Bog as well as vari[ous] wooded areas with old coppiced Sessile Oaks. The Wicklow Mounta[ins] themselves are granite and hold high concentrations of lead, tin, copper, i[ron] and zinc, all of which have been mined in the past. Most of Ireland's mamm[als] can be found here; various species of deer, foxes, badgers, brown and [pine] hares, red squirrels, birds of prey, red grouse and many other birds and f[lora] The National Park Information Office is near the Upper Glendalough lake[s]

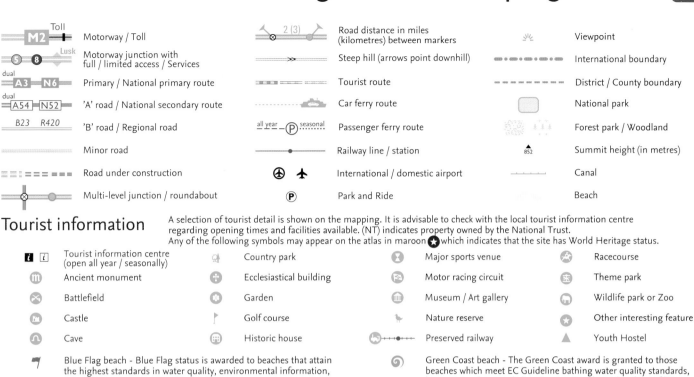

| | | |
|---|---|---|
| **M2** Toll — Motorway / Toll | 2 (3) — Road distance in miles (kilometres) between markers | Viewpoint |
| 5 8 Lusk — Motorway junction with full / limited access / Services | >> — Steep hill (arrows point downhill) | International boundary |
| dual **A3 N6** — Primary / National primary route | — Tourist route | District / County boundary |
| dual **A54 N52** — 'A' road / National secondary route | — Car ferry route | National park |
| B23 R420 — 'B' road / Regional road | all year P seasonal — Passenger ferry route | Forest park / Woodland |
| — Minor road | — Railway line / station | 852 ▲ Summit height (in metres) |
| — Road under construction | ⊕ ✈ International / domestic airport | Canal |
| ⊗ — Multi-level junction / roundabout | P Park and Ride | Beach |

## Tourist information

A selection of tourist detail is shown on the mapping. It is advisable to check with the local tourist information centre regarding opening times and facilities available. (NT) indicates property owned by the National Trust. Any of the following symbols may appear on the atlas in maroon ★ which indicates that the site has World Heritage status.

| | | | |
|---|---|---|---|
| *i* *i* Tourist information centre (open all year / seasonally) | Country park | Major sports venue | Racecourse |
| Ancient monument | Ecclesiastical building | Motor racing circuit | Theme park |
| Battlefield | Garden | Museum / Art gallery | Wildlife park or Zoo |
| Castle | Golf course | Nature reserve | Other interesting feature |
| Cave | Historic house | Preserved railway | Youth Hostel |

⚑ Blue Flag beach - Blue Flag status is awarded to beaches that attain the highest standards in water quality, environmental information, education, management and safety & services.

Only the Blue Flag symbol is shown if a beach has both Blue Flag and Green Coast awards.

Green Coast beach - The Green Coast award is granted to those beaches which meet EC Guideline bathing water quality standards, have a natural, unspoiled environment and are also clean of litter. They do not need to have the range of facilities that are required to attain blue flag status.

| water | 0 | 100 | 200 | 300 | 400 | 500 | 700 | 1000 | metres |
|---|---|---|---|---|---|---|---|---|---|
| | 0 | 330 | 650 | 980 | 1310 | 1640 | 2295 | 3280 | feet |

**SCALE**

| 0 | 2 | 4 | 6 | 8 | 10 miles |
|---|---|---|---|---|---|
| 0 | 2 4 | 6 | 8 10 | 12 14 | 16 km |

1:328 000   5.2 miles to 1 inch / 3.28 km (2 miles) to 1 cm

# Tourist information centres

## Northern Ireland
Local telephone numbers are shown. To telephone Northern Ireland from Ireland, replace 028 with 048.

| | | |
|---|---|---|
| Antrim, *Antrim & Newtownabbey*..... ☎ 028 9442 8331 **16 C4** | ✳ Bushmills, *Causeway Coast & Glens* ☎ 028 2073 0390 **16 B1** | Kilkeel, *Newry, Mourne & Down*....... ☎ 028 4176 2525 **23 D4** |
| Armagh, *Armagh City,* | Carrickfergus, *Mid & East Antrim* ... ☎ 028 9335 8262 **17 E4** | Limavady, *Causeway Coast & Glens* ☎ 028 7776 0650 **16 A2** |
| *Banbridge & Craigavon* ................ ☎ 028 3752 1800 **22 B2** | Coleraine, | Lisburn, *Lisburn & Castlereagh*......... ☎ 028 9266 0038 **23 D1** |
| Ballycastle, | *Causeway Coast & Glens* ............. ☎ 028 7034 4723 **16 B2** | Londonderry (Derry), |
| *Causeway Coast & Glens*............. ☎ 028 2076 2024 **16 C1** | Cookstown, *Mid Ulster*.................... ☎ 028 8676 9949 **22 B1** | *Derry City & Strabane*.................. ☎ 028 7126 7284 **15 F3** |
| Ballymena, *Mid & East Antrim* ........ ☎ 028 2563 5010 **16 C3** | Downpatrick, | Magherafelt, *Mid Ulster* ................. ☎ 028 7963 1510 **16 B4** |
| Ballymoney, | *Newry, Mourne & Down* ............... ☎ 028 4461 2233 **23 E2** | Newcastle, |
| *Causeway Coast & Glens*................ ☎ 028 2766 0230 **16 B2** | Dungannon, *Mid Ulster*................... ☎ 028 8772 8600 **22 A1** | *Newry, Mourne & Down*............... ☎ 0330 137 4046 **23 D3** |
| Banbridge, *Armagh City,* | Enniskillen, *Fermanagh & Omagh* ... ☎ 028 6632 3110 **21 E2** | Newry, *Newry, Mourne & Down* ....... ☎ 0330 137 4046 **22 C3** |
| *Banbridge & Craigavon* ................ ☎ 028 4062 0232 **22 C2** | George Best Belfast City Airport, | Newtownards, |
| Bangor, *Ards & North Down* ........... ☎ 028 9127 0069 **17 E4** | *Belfast City* .................................. ☎ 028 9093 5372 **23 D1** | *Ards & North Down* ..................... ☎ 028 9182 6846 **23 E1** |
| Belfast, *Belfast City* ......................... ☎ 028 9024 6609 **23 D1** | Giant's Causeway, | Omagh, *Fermanagh & Omagh*......... ☎ 028 8224 7831 **21 F1** |
| Belfast International Airport, | *Causeway Coast & Glens*.............. ☎ 028 2073 1855 **16 B1** | Strabane, *Derry City & Strabane*....... ☎ 028 7138 4444 **15 E4** |
| *Antrim & Newtownabbey*.............. ☎ 028 9448 4677 **16 C4** | Hillsborough, *Lisburn & Castlereagh* ☎ 028 9268 9717 **23 D2** | The Gobbins, *Mid & East Antrim* ..... ☎ 028 9337 2318 **17 E4** |

## Ireland
Local telephone numbers are shown. If telephoning from Great Britain or within Northern Ireland replace the initial 0 with 00 353.

| | | |
|---|---|---|
| Aran Islands, *Galway*...................... ☎ 1850 230330 **25 D4** | Dublin, O'Connell Street, *Dublin City* | Longford, *Longford* ......................... ☎ 043 334 3509 **27 E2** |
| Arklow, *Wicklow* .............................. ☎ 0402 32484 **33 F4** | (Ireland)...................................... ☎ 1850 230330 **29 E4** | Mallow, *Cork*.................................... ☎ 022 51340 **42 A1** |
| ✳ Athlone, *Westmeath* ...................... ☎ 1850 230330 **27 E3** | (UK)............................................ ☎ 0800 039 7000 **29 E4** | Midleton, *Cork*................................. ☎ 021 461 3702 **42 C2** |
| Ballina, *Mayo*................................... ☎ 096 72800 **19 F3** | Dublin Tourism Centre, | Mullingar, *Westmeath*...................... ☎ 044 934 8650 **28 B3** |
| ✳ Bantry, *Cork* .................................. ☎ 027 50229 **41 D3** | Barnardo Square, *Dublin City* | Newport, *Mayo* ............................... ☎ 098 41895 **19 D4** |
| Belmullet, *Mayo*............................... ☎ 097 20494 **18 C2** | (Ireland )...................................... ☎ 1850 230330 **29 E4** | Portlaoise, *Laois*.............................. ☎ 057 862 1178 **32 B3** |
| Blacklion, *Cavan* ............................. ☎ 071 985 3941 **21 D3** | (UK)............................................ ☎ 0800 039 7000 **29 E4** | Shannon Airport, *Clare*..................... ☎ 061 471664 **30 B4** |
| Bundoran, *Donegal* (open weekends | Duncannon (Hook Peninsula), | Skibbereen, *Cork*............................. ☎ 028 21489 **41 E4** |
| only during winter).................. ☎ 071 984 1350 **20 C2** | *Wexford*...................................... ☎ 051 387530 **38 C3** | Slane, *Meath* ................................... ☎ 041 982 4000 **29 D2** |
| ✳ Cahir, *Tipperary*............................. ☎ 052 7741453 **37 F3** | Dundalk, *Louth* ............................... ☎ 042 935 2111 **22 C4** | Sligo, *Sligo* ...................................... ☎ 071 916 1201 **20 B3** |
| Carlow, *Carlow* ................................ ☎ 059 913 0441 **32 C4** | Dungarvan, *Waterford*...................... ☎ 058 41741 **38 A4** | Tralee, *Kerry* .................................... ☎ 066 712 1288 **35 D4** |
| ✳ Carrick-on-Shannon, *Leitrim* ......... ☎ 071 962 0170 **27 D1** | Dungloe, *Donegal* ........................... ☎ 074 952 2198 **14 B3** | Trim, *Meath* ..................................... ☎ 046 943 7227 **29 D3** |
| Cavan, *Cavan* .................................. ☎ 049 433 1942 **21 F4** | Dún Laoghaire, | Tubbercurry (Tobercurry), *Sligo*........ ☎ 087 093 6616 **20 A4** |
| Claremorris, *Mayo*........................... ☎ 094 937 1830 **26 A2** | *Dún Laoghaire-Rathdown* ............. ☎ 01 280 6964 **33 F1** | Waterford, *Waterford* ....................... ☎ 051 875823 **38 C3** |
| ✳ Clifden, *Galway*.............................. ☎ 095 21163 **24 C2** | Ennis, *Clare*..................................... ☎ 1850 230330 **30 B4** | Westport, *Mayo*................................ ☎ 1850 230330 **19 D4** |
| Clonakilty, *Cork*............................... ☎ 023 883 3226 **41 F3** | Fermoy, *Cork*................................... ☎ 025 82886 **42 C1** | Wexford, *Wexford*............................ ☎ 1850 230330 **39 E2** |
| Clonmel, *Tipperary*........................... ☎ 052 612 2960 **38 A2** | Galway, *Galway City* ........................ ☎ 1850 230330 **26 A4** | Wicklow, *Wicklow*............................ ☎ 0404 69117 **33 F3** |
| Cobh, *Cork* ...................................... ☎ 021 481 3301 **42 B2** | Glen of Aherlow (Newton), | Youghal, *Cork* .................................. ☎ 024 20170 **43 D2** |
| ✳ Cong, *Mayo*................................... ☎ 094 954 6542 **25 E2** | *Tipperary*.................................... ☎ 062 56331 **37 E3** | |
| Cork, *Cork City*................................. ☎ 1850 230330 **42 B2** | ✳ Kenmare, *Kerry* ............................ ☎ 064 664 1233 **41 D2** | **For further information:** |
| Dingle, *Kerry* ................................... ☎ 066 915 1188 **34 B4** | Kildare, *Kildare*................................ ☎ 045 530672 **32 C2** | |
| Donegal, *Donegal*............................. ☎ 1850 230330 **20 C1** | Kilkenny, *Kilkenny* ........................... ☎ 1850 230330 **38 B1** | **Northern Ireland** |
| Drogheda, *Louth* .............................. ☎ 041 987 2843 **29 E2** | Killarney, *Kerry* ............................... ☎ 1850 230330 **41 D1** | www.discovernorthernireland.com |
| Dublin Airport, *Fingal* | Kinsale, *Cork* ................................... ☎ 021 477 2234 **42 B3** | **Ireland** |
| (Ireland)...................................... ☎ 1850 230330 **29 E3** | Letterkenny, *Donegal* ...................... ☎ 1850 230330 **15 D3** | www.discoverireland.ie |
| (UK) ........................................... ☎ 0800 039 7000 **29 E3** | Limerick, *Limerick* ........................... ☎ 061 317522 **36 C2** | |
| | Listowel, *Kerry*................................ ☎ 068 22212 **35 D3** | ✳ Seasonal opening. |

**A** **B** **C**

**1**

**2**

**3**

**4**

0 2 4 6 miles
0 2 4 6 8 10 km

Tory Island

DIXON GALLERY

An Baile Thiar
(West Town)

Tory    Sound

Inishbeg

THE WOR

Inishdooey

Inishbofin

Dooras Pt

Ranaghroe
Point

Bloody Foreland

Bloody
Foreland

Min Larach
(Meenlaragh)

R251

An Fál
(Falcarra

Bun na Leaca
(Brinlack)

314

Min an Chladaigh
(Meenaclady)

Inishsirrer

Gort an Choi
(Gortahork)

Inishmeane

Gweedore

Lough
Lagha

Gola Island

Tievealehid

Cloghaneel

Doirí Beaga
(Derrybeg)

430

N56

Aghla More

An Bun Beag
(Bunbeg)

Gaoth Dobhair
(Gweedore)

584

Owey I.

Donegal

R258

Altan
Lough

Cruit I.

Inishfree
Bay

R266

Errigal

Rosses Bay

Cionn Caslach
(Kincaslough)

Croithlí
(Crolly)

36
(58)

Lough
Nacung
Upper

752

Torneady
Point

Anagaire
(Annagry)

IONAD COIS LOCHA/
THE DUNLEWEY CENTRE

Dún Lúiche
(Dunlewy)

Rinrawros Point

The Rosses

Crocknafarragh

Derryveagh

An Leadhb Gharbh
(Leabgarrow)

517

Slieve
Snaght

Arranmore
Island

228

R259

Ailt an Chorráin
(Burtonport)

Loch an Iúir
(Loughanure)

683

Moylena

Rutland
Island

Lough
Anure

540

Inishfree Upper

Lough
Meela

Crovehy

Lough Barra

D O N

Maghery Bay

An Clochán Liath
(Dungloe)

316

i

Lough
Craghy

R252

R254

Lo

An Machaire
(Maghery)

Dunglow
Lough

Lough
Neck
More

Crohy Head

Doire Eadarúil
(Derrydruel)

N56

Lough
Gannivegil

An Dúchoraidh
(Doochary)

Baile na
Finne
(Fintown)

R251

Tie

25

Mín na Croise
(Meenacross)

23
(37)

Croaghleheen

Gweebarra Bay

Dooey
Point

383

FINTOWN
RAILWAY

Mín a
(Mee

Dunmore
Head

Toome
Lough

Ballynacarrick

Aghla
Mountain

Lough Finn

Portnoo

Narin

Clooney

Léitir Mhic an Bhaird
(Lettermacaward)

R250

Dawros
Head

KILCLOONEY
DOLMEN

Maas

598

An Ghrafaidh
(Graffy)

R253

Rossbeg

Kilclooney

ST. CONNELL'S
MUSEUM

An tSeanga Mheáin
(Tangaveane)

Loughros
More Bay

R261

Lough
Machugh

Lough
Ananima

Glenties

Loughros Point

Kilrean

Lough
Anna

Cloghboy

Crannogeboy

MAGHERA CAVES

Film Location:
DANCING AT LUGHNASA

Blue
Stack

Meenacurrin

An Machaire
(Maghera)

Glengesh
Hill

Ardara
(Ard an Rátha)

Tullyhonwar

Carnaween

676

Lough
Anaffrin

511

385

522

Blue Stack Mts

Stravally
(Srath an Bhealaigh)

Lough
Nalughraman

Binbane

FATHER McDYER'S
FOLK VILLAGE

OIDEAS
GAEL

374

Glengesh Pass

N56

455

R262

Glen
Bay

Crocknapeast

Tamor
Lough

Rossan Pt.

Gleann Cholm Cille
(Glencolumbkille)

502

Meentullynagarn

Letterbarra

Málainn Mhóir
(Malin More)

305

Min an Aoire
(Meenaneary)

Lough
Adeery

Olly

Croagh

Drumnalost

Malin
Bay

Lough
Unna

Corker

Rathlin
O'Birne I.

An Charraig
(Carrick)

Frosses

Lough
Ess

Málainn Bhig
(Malin Beg)

Slieve
League

Crownarad

494

Calhame

Inver

Donegal
(Dún na nGall)

i

601

SLIEVE LEAGUE CLIFFS CENTRE

Cill
Charthaigh
(Kilcar)

25
(40)

Bruckless

Mountcharles

R26

Teileann (Teelin)

R263

Killybegs
(Na Cealla Beaga)

20

SALTHILL
GARDENS

Lagh

Carrigan Head

Muckross
Head

Muckross
(Mucros)

Drumanoo

Tullyvoos

**A** **B** **C**

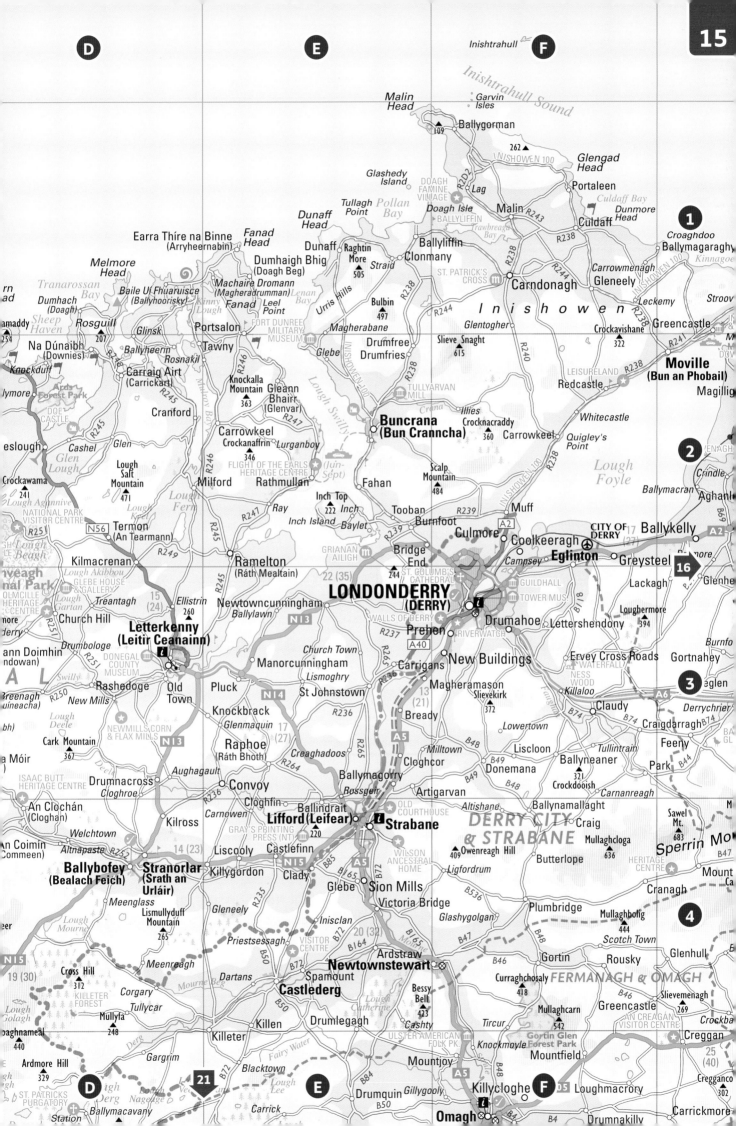

Bennan Head

Machrihanish Bay

Davaar Island

**D**

Machrihanish

A83

**E** Campbeltown

**F**

B843

Cnoc Moy 446

Conie Glen

B842

Feochaig

Macharioch

Southend

🚗 ARDROSSAN (May-Sept)

| 0 | 2 | 4 | 6 miles |
| 0 | 2 | 4 | 6 | 8 | 10 km |

Mull of Kintyre

Sanda Island

NORTH CHANNEL

Ⓟ

**1**

Ailsa Craig

-CRANNAGH CRANNOG

gh Bay

Torr Head

Loughan Bay

27)

Tornamoney Point

B92

Cushendun Bay

Cushendun

Knocknacarry

★ CURFEW TOWER

✝ LAYDE CHURCH

Cushendall Bay

Cushendall

f

Red Bay

Garron Point

**2**

Milleur Point

10 (16)

Knockore 359

B738

A7

Kirkcolm

B198

B738

Leswalt

B7043

TOR CENTRE

3

BELFAST PLACES OF INTEREST
① BARNETT DEMESNE
② BELFAST CITY HALL
③ BELVOIR PARK FOREST
④ BOTANIC GARDENS & ULSTER MUSEUM
⑤ GIANT'S RING
⑥ GRAND OPERA HOUSE
⑦ TITANIC BELFAST

**3**

Portpatrick

Hunters Point

Carnlough

Carnlough Bay

A42

Straidkilly

Glenarm

❀ GLENARM CASTLE WALLED GARDEN

15 (24)

Slane

B97

Black Hill 383

A2

Feystown

The Sheddings

The Maidens or Hulin Rocks

🚗

CAIRNRYAN

MID & ST ANTRIM

Ballygalley

Ballygalley Head

CARNFUNNOCK

Drains Bay

Ballycraigy

Glen Head 393

Agnew's Hill 476

Carncastle

Sallagh Braes

B148

Skernaghan Point

Isle of Muck

CAIRNRYAN

Millbrook

Larne

A8

Mullaghboy

Islandmagee

Shoptown

Kilwaughter

Glynn

Millbay

Magheramorne

⊗ KILCOAN GARDENS

A36

22 (35)

B100

Raloo

Glenoe

B99

B149

20 (33)

B50

ℹ THE GOBBINS CLIFF PATH

Killylane Reservoir

A8

11 (18)

Slimero 291

Ballycarry

Lough Mourne

B90

Ballystrudder

Black Head

Tildarg

Ballynure

Ballyeaston

nside

Ballynure

Straid

B58

LIVERPOOL (BIRKENHEAD) DOUGLAS (Apr-Sept)

🚗

**4**

Ballyclare

A8

Carn Hill 312

Whitehead

ll Doagh

A57

B95

A8

B58

Belmont

ANDREW JACKSON COTTAGE & US RANGERS CENTRE

Light House Island

Mew Island

Ballyrobert

Ⓟ

Eden

ℹ

🏛 CARRICKFERGUS MUS

CRAWFORDSBURN

4 (6) Templepatrick

patrick

1 (2)

A8(M)

Mossley

B90

A2

**CARRICKFERGUS**

Greenisland

Helen's Bay

Grey Pt

🏛 CRAWFORDSBURN

Groomsport

Copeland Island

SON'S L (NT)

Mallusk

Glengormley

B59

TRANSPORT MUS.

B20

ℹ Ballyholme

NORTH DOWN MUSEUM

NTOWNABBEY

oanends

ZOO 🎡

Belfast Lough

A2

🏛 FOLK MUS.

3 21

**BANGOR**

BELFAST CASTLE

CAVE HILL

M5

Holywood

SOMME MUSEUM

Conlig A48

Six Road Ends

ℹ

★ BALLYCOPELAND WINDMILL

Ligoniel

Corner

BEL ST C

Ⓟ

7

BELFAST CITY

ℹ ✈

Craigantlet

REDBURN

SSE ARENA BELFAST

MONT

THE ARK OPEN FARM

Donaghadee

A2

Millisle

**BELF** D T

ahstown

938

✝ Ⓟ

2

ℹ

Ⓟ

23 ⬇

4 RISE

A55

20

**Dundonald**

9 (14)

STREAMVALE OPEN FM

SCRABO

**E** NEWTOWNARDS

Ballywatticock

**F**

Drumfad

B172

Ballyferis Point

ARDS or NORTH DOWN

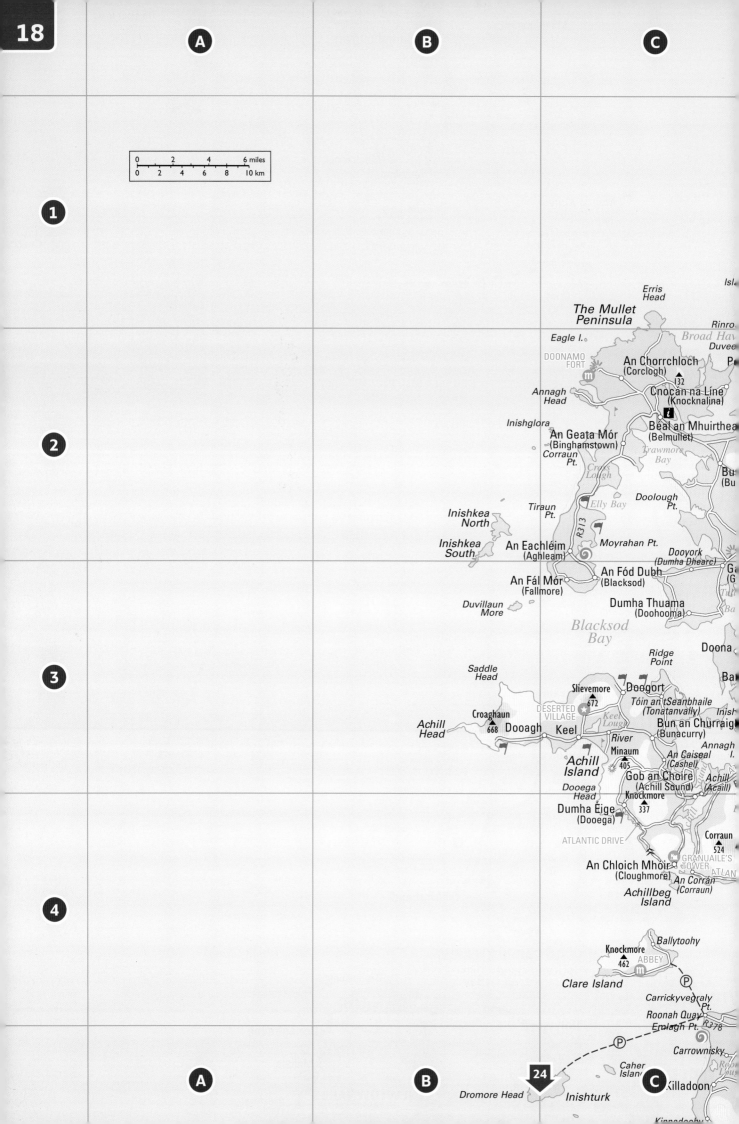

A

B

C

0  2  4  6 miles
0  2  4  6  8  10 km

1

2

3

4

Erris Head

The Mullet Peninsula

Isl

Rinro

Eagle I.

Broad Hav

Duvee

DOONAMO FORT

An Chorrchloch (Corclogh)

P

132

Annagh Head

Cnocán na Líne (Knocknalina)

Inishglora

Béal an Mhuirthea (Belmullet)

An Geata Mór (Binghamstown)

Trawmore Bay

Corraun Pt.

Bu (Bu

Cross Lough

Elly Bay

Doolough Pt.

Inishkea North

Tiraun Pt.

R313

Inishkea South

An Eachléim (Aghleam)

Moyrahan Pt.

Dooyork (Dumha Dhearc)

An Fál Mór (Fallmore)

An Fód Dubh (Blacksod)

Ga (G

Duvillaun More

Dumha Thuama (Doohooma)

Ba

Blacksod Bay

Doona

Ridge Point

Saddle Head

Slievemore 672

Doogort

Ba

Croaghaun 668

DESERTED VILLAGE

Tóin an tSeanbhaile (Tonatanvally)

Inish

Achill Head

Dooagh

Keel

Keel Lough

Bun an Churraig (Bunacurry)

River Minaum 405

Annagh I.

Achill Island

An Caiseal (Cashel)

Dooega Head

Gob an Choire (Achill Sound)

Achill (Acaill)

Dumha Éige (Dooega)

Knockmore 337

ATLANTIC DRIVE

Corraun 524

GRANUAILE'S TOWER

An Chloich Mhór (Cloughmore)

ATLAN

An Corrán (Corraun)

Achillbeg Island

Knockmore 462

Ballytoohy

ABBEY

Clare Island

Carrickyvegraly Pt.

Roonah Quay

Enlagh Pt.

R378

24

Carrownisky

Caher Island

Roo Loug

Dromore Head

Inishturk

Killadoon

Kinnadoohy

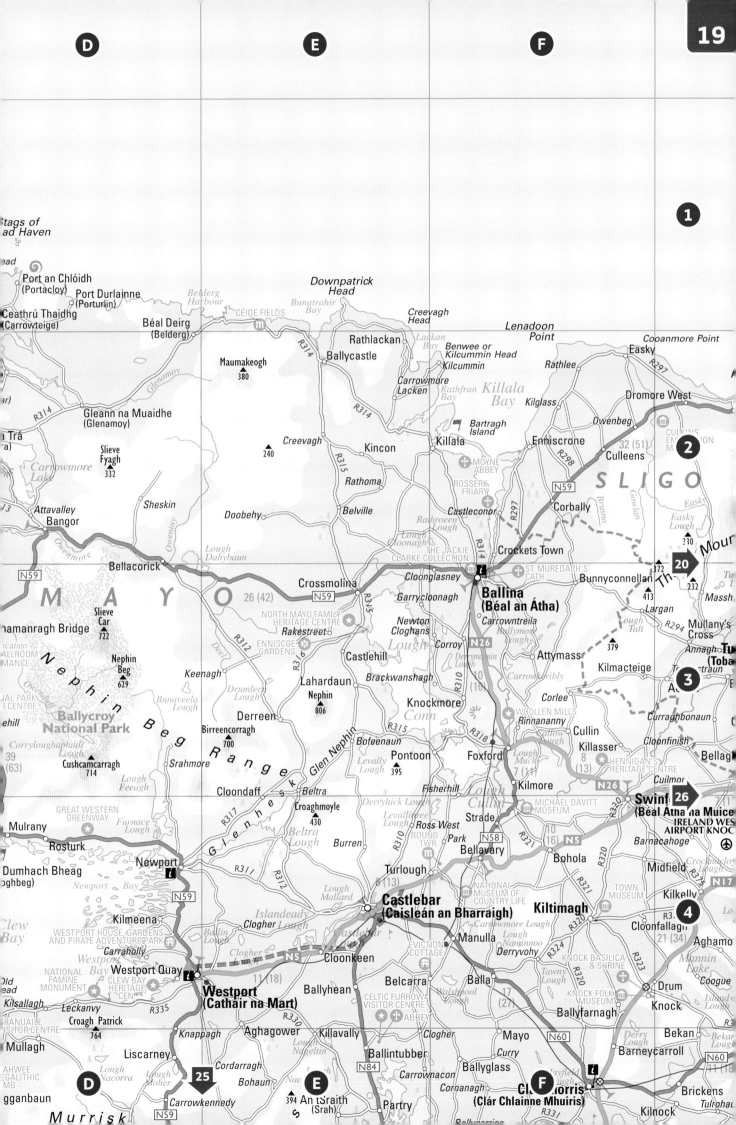

D     E     F

1

2

3

4

20

26

25

*Stags of*
*ad Haven*

*Port an Chlóidh*
*(Portacloy)*

*Port Durlainne*
*(Porturlin)*

*Ceathrú Thaidhg*
*(Carrowteige)*

*Béal Deirg*
*(Belderg)*

*Belderg*
*Harbour*

CÉIDE FIELDS

*Bunatrahir*
*Bay*

*Downpatrick*
*Head*

*Creevagh*
*Head*

Rathlackan

Ballycastle

*Carrowmore*
*Lacken*

*Lacken*
*Bay*

Benwee or
Kilcummin Head

Kilcummin

*Rathfran*
*Bay*

*Killala*
*Bay*

Rathlee

Rathlackan

*Lenadoon*
*Point*

*Cooanmore Point*

Easky

Dromore West

R297

Gleann na Muaidhe
(Glenamoy)

*Glenamoy*

*Trá*

R314

Attavalley

Bangor

*Carrowmore*
*Lake*

Slieve
Fyagh
▲332

Sheskin

Maumakeogh
▲380

R314

*Creevagh*

Kincon

▲240

Killala

Kilglass

*Bartragh*
*Island*

Enniscrone

Owenbeg

Cullens

R298

Corbally

*Brusna*

*Gowlan*

*Easky*

*Easky*
*Lough*

▲330

▲372

▲232

CULKINS
EMIGRATION
M

SLIGO

Doobehy

Belville

Rathoma

MOYNE
ABBEY

ROSSERK
FRIARY

Castleconor

R314

R297

Crockets Town

ST. MUREDACH'S
CATH.

▲413

Bunnyconnellan

*Th*

*Mour*

Massh

N59

N59

Bellacorick

MAYO

Crossmolina

26 (42)

N59

R315

Cloonglasney

THE JACKIE
CLARKE COLLECTION

Ballina
(Béal an Átha)

Garrycloonagh

Newton
Cloghans

Carrowntreila

R294

▲379

Largan

*Lough*
*Talt*

Mullany's
Cross

Annagh

Tu
(Toba
traun)

*Owenmore*

Slieve
Car
▲722

Rakestreet

NORTH MAYO FAMILY
HERITAGE CENTRE

ENNISCOE
GARDENS

R316

Castlehill

R312

*Deel*

Keenagh

*Drumlee*
*Lough*

Lahardaun

Nephin
▲806

Brackwanshagh

Corroy

N26

Derrynanin
Lough

Attymass

Kilmacteige

Kilmacteige

▲379

R310

Corlee

Cullin

Killasser

Cloonfinish

Bellag

*Currughbonaun*

Slieve
Beg
▲629

Derreen

Birreencorragh
▲700

*Bunaveela*
*Lough*

Srahmore

Bofeenaun

*Levally*
*Lough*

Pontoon

Nephin
▲395

Knockmore

*Lough*
*Conn*

R315

Rinnananny

WOOLLEN MILLS

HENNIGAN'S
HERITAGE CENTRE

8
(13)

Cuilmor

Swinf
(Béal Átha na Muice)

Cushcamcarragh
▲714

*Lough*
*Feeagh*

Cloondaff

*Glenhest*

*Glen Nephin*

Beltra

Croaghmoyle
▲430

Fisherhill

Foxford

*Lough*
*Muck*

Kilmore

7
(11)

N26

Michael Davitt
Museum

IRELAND WES
AIRPORT KNOC

*Great Western*
*Greenway*

Mulrany

*Furnace*
*Lough*

Rosturk

Dumhach Bheag
*(oghbeg)*

Newport

t

*Newport*
*Bay*

N59

Kilmeena

Carraholly

WESTPORT HOUSE, GARDENS
AND PIRATE ADVENTURE PARK

*Clew*
*Bay*

*Old*
*ead*

*Kilsallagh*

Leckanvy

*RANUAILE*
*VISITOR CENTRE*

NATIONAL
FAMINE
MONUMENT

CLEW BAY
HERITAGE
CEN

Croagh Patrick
▲764

Westport Quay

t

Westport
(Cathair na Mart)

R335

Knappagh

Aghagower

Killavally

R330

Liscarney

Cordarragh

Bohaun

Carrowkennedy

*Murrisk*

▲394 An tSraith
(Srah)

*S*

Partry

R317

*Ballin*
*Lough*

R311

*Islandeady*

Clogher

Clogher *Lough*

*Ballin*
*Lough*

*Clogher*

Burren

Ross West
Park

*Derryhick Lough*

*Levallinree*
*Lough*

ROUND
TWR.

Strade

N58

R310

Bellavary

Turlough

8
(13)

R312

*Lough*
*Mallard*

N5

Castlebar
(Caisleán an Bharraigh)

*Castlebar*

Cloonkeen

EVICTION
COTTAGE

Manulla

Derryvohy

*Lough*
*Naminoo*

R324

*Tawny*
*Lough*

NATIONAL
MUSEUM OF
COUNTRY LIFE

N58

10
(16)

N5

Bohola

R320

R321

Barnacahoge

TOWN
MUSEUM

Midfield

*Croc*

N17

R375

Kilkelly

Cloonfallagh

21 (34)

Aghamo

*Mannin*
*Lake*

Coogue

*11 (18)*

Cloonkeen

Belcarra

Balla

WALSHPOOL
*Lough*

17
(27)

KNOCK BASILICA
& SHRINE

R320

KNOCK FOLK
MUSEUM

Drum

*Island*
*Lough*

N5

Kiltimagh

Knock

Ballyhean

Ballintubber

N84

Carrownacon

Mayo

CELTIC FURROW
VISITOR CENTRE

ABBEY

*Clogher*

Curry

Cornanagh

Ballyglass

N60

*Bayfield*
*Lough*

t

Cl. orris
(Clár Chlainne Mhuiris)

Ballyfarnagh

Bekan

Barneycarroll

*Derry*
*Lough*

*Bekar*
*Lough*

N60

11(1

Brickens

Tulroha

Kilnock

R331

Kilmeena

Clew
Bay

Dumhach Bheag

*Ballygarries*

Partry

▲394

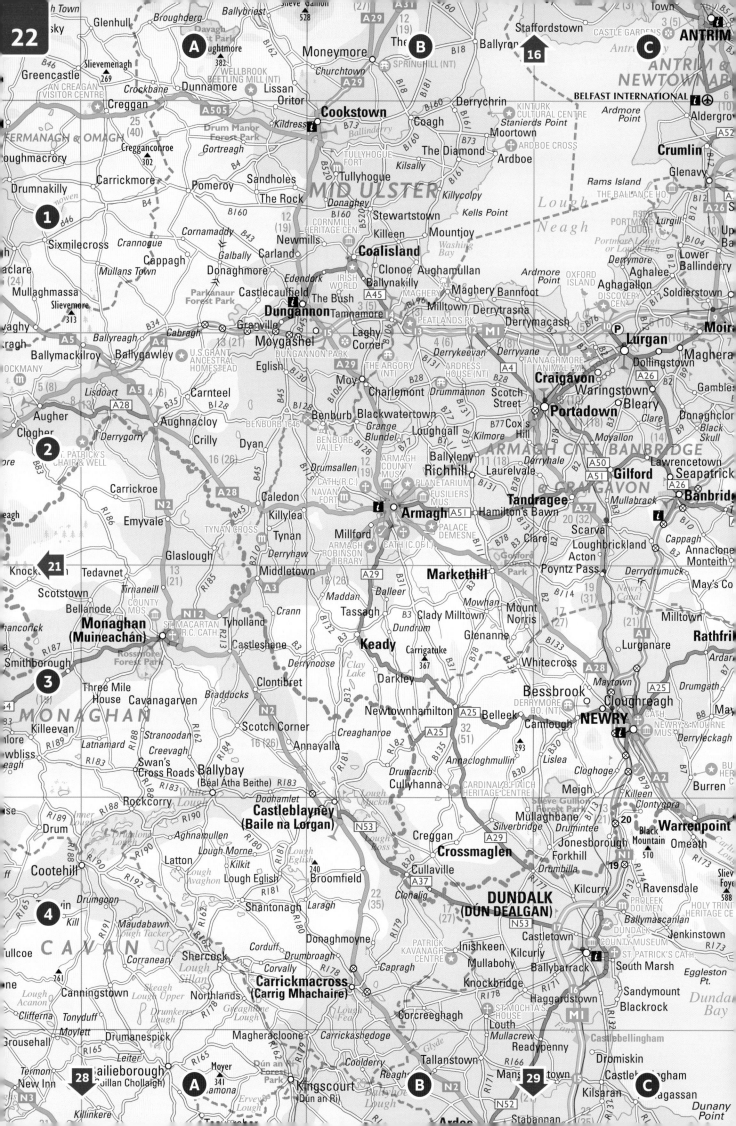

## BELFAST PLACES OF INTEREST

1. BARNETT DEMESNE COUNTRY PARK
2. BELFAST CITY HALL
3. BELVOIR PARK FOREST
4. GIANT'S RING
5. GRAND OPERA HOUSE
6. NATIONAL BOTANIC GARDENS
7. TITANIC BELFAST

A

B

18

C

Knockmore
462

Ballytoohy

ABBEY

Clare Island

Carrickvegraly Pt.

Roonah Quay

Emlagh Pt.

R378

Carrownisky

Caher
Island

Killadoon

Dromore Head

Inishturk

Kinnadoohy

Tonakeera
Point

1

Inishbofin

Bunnamullen
Bay

Crump
Island

Rinvyle
Point

Salroc

Inishark

Bofin
Harbour

Inishlyon

Ardnagreevagh

Cashleen

Rinvyle

Gowlaun

Tully
Lough

Tully Cross

60

Aughrusbeg
Lough

Cleggan
Bay

OCEAN & COUNTRY
VISITOR CENTRE

Letterfrack

NATIONAL PARK
VISITOR CENTRE

KY
KY

Aughrus More

Cleggan

Ballynakill
Lough

Moyard

Connemara
National Park

High I.

Claddaghduff

Cruagh

Omey
I.

Inishturk

N59

Lough
Auna

Lough
Nahillion

CONNEMARA
HERITAGE &
HISTORY CE

Turbot I.

CLIFDEN
CASTLE

i

Clifden

Derrylea

2

ALCOCK & BROWN
MONUMENT

Mannin
Bay

Derrylea
Lough

Ballinaboy

Ballynahinch

Bally
L

Doonloughan

R34

Ballyconneely

Toombe

Slyne Head

CONNEMARA
SMOKEHOUSE

Ballyconneely
Bay

Callow

Errisbeg
300

Lough
Bollard

Inishnee

Roundston

Inishlackan

Bertraghbo

Mairos
(Moyrus)

3

Croaghnakeela
Island

Mace
Head

An Aird
(Ard)

St Macdara's
Island

Mweenish
Island

4

Eog

Brannock I.

DUN A
(DUN

A

B

C

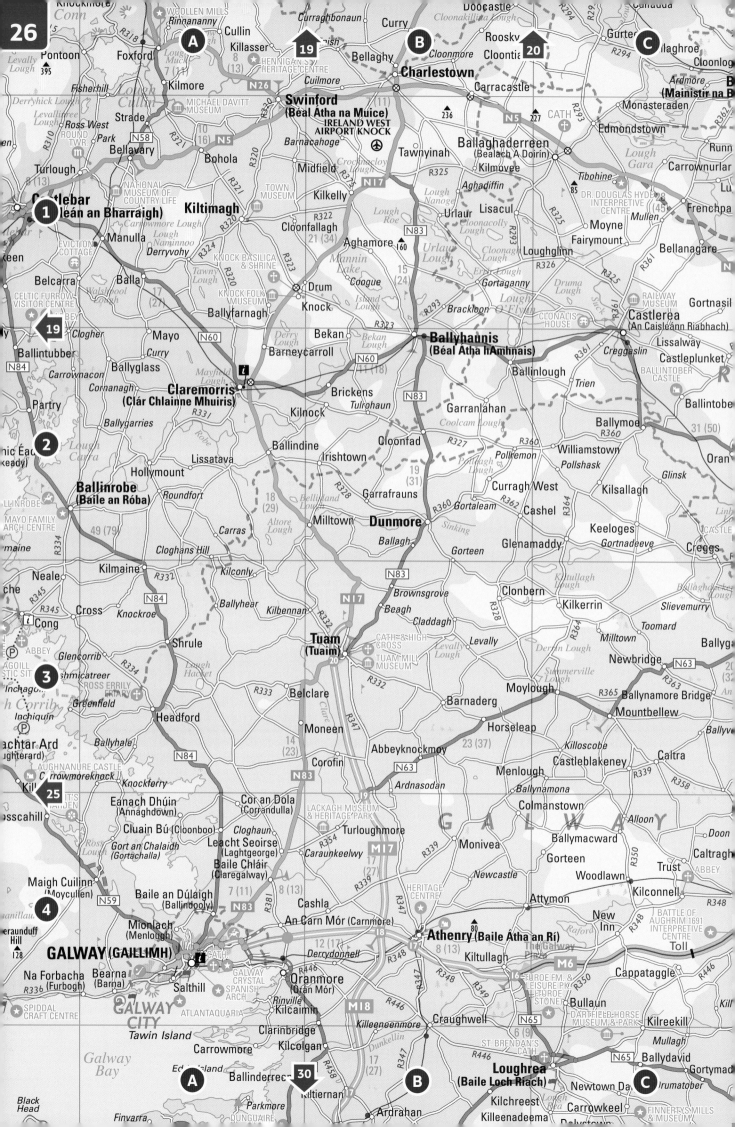

DUBLIN PLACES OF INTEREST
1 BOOK OF KELLS
2 CHRIST CHURCH CATHEDRAL
3 CROKE PARK, GAA MUSEUM & CROKE PARK SKYLINE
4 DUBLIN CASTLE
5 DUBLINIA
6 GUINNESS STOREHOUSE
7 IRISH MUSEUM OF MODERN ART
8 JAMESON DISTILLERY
9 ST. MARY'S CATHEDRAL
10 ST. PATRICK'S CATHEDRAL

The M50 is now barrier-free between junctions 6 and 7. Unregistered users can pay the toll in shops or garages where a 'Payzone' logo is displayed. Alternatively the toll can be paid online at www.eflow.ie or by calling 1890 50 10 50. The toll must be paid by 8pm on the following day.

M50 Dublin Port Tunnel: Toll for non-HGV traffic only

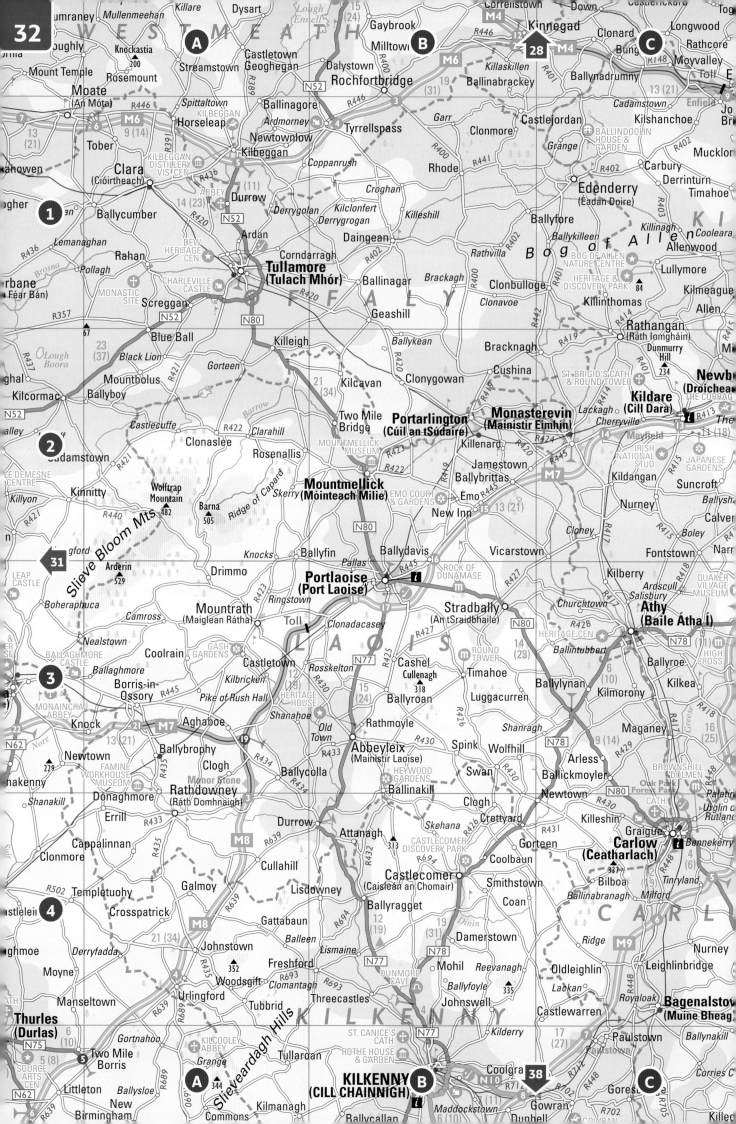

A    B    C

1

ATLANTIC

OCEAN

2

*Tullig Point*

Feeard

MONEEN CHURCH

*Loop Head*    78    Kilbaha

*Dunmore Head*    Kilclo Po

(Baile

3

*Inshaboy Point*

Dreenagh    Maulin Mountain

*Kerry Head*    218

Glenderry

Ballyheigue

*Ballyheigue Bay*

Banna

CATH

*The Seven Hogs*    *Rough Pt.*

4

Brandon Pt.    Fahamore    Chapeltown

*Brandon Head*    *Brandon Bay*    *Tralee Bay*    Fenit

Cé Bhréanainn (Brandon)    Caher Pt.    Lough Gill    Castlegregory    *Derrymore Island*    Ble

*Ballydavid Hd.*    Brandon Mountain    Stradbally

An Fheothanach (Feohanagh)    R549    953    R560    Kilcummin    Aughacasla    Derryn

*Sybil Hd.*    *Smerwick Harbour*    An Clochán (Cloghane)    *D i n g l e*    R560    Camp    Baurtregaum

Ard na Caithne (Smerwick)    An Mhuiríoch (Murreagh)    Baile an Lochaigh (Ballinloghig)    Beenoskee    852

*Sybil Pt.*    Ballysitteragh    827    32 (51)    N86    *S l i e v e*

624    Road unsuitable for HGV's and buses    Moanlaur    Fybagh

Baile an Fheirtéaraigh (Ballyferriter)    GALLARUS ORATORY    Cill Maoilchéadair (Kilmalkedar)    Conor Pass    Knockmulanane    566    *Lougher*

594    Aughils Bridge

Film Location: RYAN'S DAUGHTER    Anascaul

Baile an Mhuilinn (Milltown)    OCEANWORLD    R561    Inch

*Inishtooskert*    THE BLASKET CENTRE    Dún Chaoin (Dunquin)    Daingean Uí Chúis (Dingle)    Lios Póil (Lispole)    *Castlemaine Harbour*    Tullig

Mt. Eagle    Ceann Trá (Ventry)    Knockaunnaglashy    Cromar

*Great Blasket I.*    516    R559    Inch

*Tearaght Island*    Croaghmore    *Parkmore Point*    Minard Head    40    sbehy Point    KERRY B VILLAGE M    *Lough Yganaven*

292    A    Slea Head    B    Bay    C

*Inishnabro*    N70

Scale: 0 2 4 6 miles / 0 2 4 6 8 10 km

Ardattin
419
Ballard
Coolattin
Inch
Kilmichael
Point

N81
ALTAMONT
GARD
33
Killinierin
Gorey
Castletown
F

D
Derry Water
Carnew
Mona
E
Hollyfort
R772
Clonegal
Knockbrandon
22
Tara
Hill
253
M11

R724
N80
R746
Gorey (Guaire)
KIA ORA
MINI FARM

Kildavin
Askamore
Craanford
R725
R772
23

19
(31)
315
Slieveboy
422
Clogh
25
(42)
Courtown
1

Bunclody
(Bun Clóidi)
Ballyroebuck
Riverchapel

Kilmyshall
Clohamon
Ballyduff
Camolin
24
Ballycanew
Roney
Point

Strahart
Balloughter
Killenagh

R746
Tombrack
R772
Ferns
Ballygarrett
R742

Templeshambo
Ballycarney
W E X F O R D
Milltown
ST. EDAN'S
CATH.
BALLYMORE
HISTORIC
MUSEUM
Clonevin
Cahore Point

Ballindaggan
The
Harrow
Boolavogue
Monamolin

Kiltealy
R102
Marshalstown
Mile
House
Ballysimon
WELLS HOUSE & GARDENS

R730
ST. AIDAN'S
CATH.
25
Oulart
Kilnamanagh
Kilmuckridge

R730
ENNISCORTHY
CASTLE
Enniscorthy
(Inis Córthaidh)
Kilcotty
R744

Davidstown
Film Location:
BROOKLYN
Ballaghkeen
Castlellis
R745
Ballyvaldon

COOLAUGHT
GARDENS
NATIONAL 1798
REBELLION CENTRE
Glenbrien
R741
Blackwater

23
(37)
Clonroche
Bree
Ballymurn
R742

N30
Ballyhoge
N11
Oilgate
Redgate
Screen
Wexford
Bay

The Leap
R735
Galbally
Crossabeg
R742
Film Location:
SAVING PRIVATE RYAN
2

Adamstown
14
(23)
Killurin
Curracloe

Raheen
181
Camaross
WESTGATE
TWR.
Castlebridge

bola
17
(28)
IRISH NATIONAL
HERITAGE PARK
Ferrycarrig
R741
SELSKAR
ABBEY
WEXFORD
WILDFOWL RESERVE

N25
Cullenstown
WEXFORD
Wexford
(Loch Garman)
The Raven Point

Newbawn
Taghmon
R738
R733
Wexford
Harbour
Rosslare Point

Foulkesmill
N25
Drinagh

R736
JOHNSTOWN CASTLE
GARDENS
Burrow
Rosslare
Bay

lane
Waddingtown
IRISH
AGRICULTURAL
MUS
Rosslare

Hilltown
R733
Murntown
13
(21)
3

cullen
Tullycanna
Cleristown
R740
Rosslare Harbour

Wellingtonbridge
Walshestown
Mayglass
Killinick
Tagoat
Greenore
Point
FISHGUARD
PEMBROKE DOCK

Carrick
R736
Ballycogly
Kilrane
Broadway

Duncormick
Bridgetown
Park
R736
Tacumshane
Lady's Island
CHERBOURG (Mar-Dec)
ROSCOFF (May-Sept)

Cullenstown
R739
Tomhaggard
Lady's I.

Bannow
Kilmore
TACUMSHANE
WINDMILL
Churchtown

Ballyteige
Bay
Millroad
Tacumshin
Lake
Lady's
Island Lake
Carnsore Point

Crossfarnoge or
Forlorn Point
Kilmore Quay

Saltee
Island Little

Saltee
Island Great
Saltee
Islands

0    2    4    6 miles
0  2  4  6  8  10 km

D
E
F

Ard na Caithne
(Smerwick)
Sybil Pt.
An Mhuírioch
(Murreagh)
Ballysitteragh
624
Ballinnogfing
R560
N86
32 (51)
852
Moa
Baile a 'rtéaraigh
**A**
(Ballyferriter)
GALLARUS
ORATORY
Cill Maoilchéadair
(Kilmalkedar)
Co
**B**
Knockmulanane
594
Road unsuitable for
HGV's and buses
**34**
Fybagh
Tougher
N86
Aughils Bridge
**C**
Slieve

Inishtooskert
THE BLASKET
CENTRE
Dún Chaoin
(Dunquin)
Film Location:
RYAN'S DAUGHTER
OCEANWORLD
Daingean
Uí Chúis
(Dingle)
Lios Póil
(Lispole)
R561
Inch
Castlemaine
Harbour

Great
Blasket I.
Tearaght
Island
Croaghmore
292
Mt. Eagle
516
Ceann Trá
(Ventry)
Parkmore
Point
Bull's
Head
Minard
Head
Knockaunnaglashy
Lough
Yganavan
Cromane
KERRY BOG
VILLAGE MUS.
Tullig

Inishnabro
R559
Slea
Head
Dingle
Bay
Rosbehy
Point
R564
N70
Glenbeigh
Seefin
494
Lough
Caragh

**1**
Inishvickillane
Darby's
Bridge
Kells
35
(56)
Been
Hill
670
Ballynakilly
Shanacashel
Lough
Acoose

It is advised to travel
in an anti-clockwise
direction around the
'Ring of Kerry'
Canglass
Point
Knocknadober
691
RING OF KERRY
Coomacarrea
774
Coomasaharn Lough
Bealalaw Bridge
Iveragh

Doulus
Head
Beginish I.
BALLYCARBERY
CASTLE
Cahirsiveen
(Cathair Saidhbhín)
OLD BARRACKS
HERITAGE CENTRE
Teeromoyle
Colly
688
Cloon
Lough
Mullaghanattin
774
Macgill

Beenakryraka Hd.
Valencia Island
(Valentia Island)
Beenaniller Head
Bray Hd.
Knights Town
Chapeltown
Aghnagar
Bridge
R565
R565
Foilclogh
500
Droichead Lios an tSonnaigh
(Lissatinnig Bridge)
Lough
Reagh

Clynacartan
SKELLIG
EXPERIENCE
Portmagee
Kilkeaveragh
371
An Trianlarach
(Teeranearagh)
R566
R567
An Chillín Liath
(Killeenleagh)
Derriana
Lough
Knocknagul
415

**2**
Puffin
Island
Cill Ón Chatha
(Killonecaha)
St Finan's
Bay
Dún Géagáin
(Dungeagan)
R566
Máistir Gaoithe
(Mastergeehy)
Sallahig
Coomcallee
674
Lough
Currane
KERRY GEOPARK
Sneem
N70

SKELLIG
MICHAEL
MONASTERY
The Skelligs
SCEILG MHICHÍL
Bolus
410
Ducalla
Head
Bolus Head
Baltinskelligs
Bay
Baile an Sceilg
(Ballinskelligs)
Hogs Head
Waterville
An Baile Breac
(Ballybrack)
N70
Cahernageeha
499
STAIGUE
STONE
FORT
39
(63)
Parknasilla
Tahilla
River
R573
Buna

Film Location:
STAR WARS EPISODE VII:
THE FORCE AWAKENS
DERRYNANE
HOUSE
Abbey
I.
Film Location:
EXCALIBUR
Cathair Dónall
(Catherdaniel)
Castlecove
Kenmare
DERREEN
GARDEN

Scariff I.
Deenish I.
Lamb's
Head
Kilcatherine Pt.
Collorus
Ardgroom
Glanmore
Lake

**3**
Inishfarnard
Coulagh
Bay
Eyeries
Glenbeg
Lough
Maulin
622
Be
Hungry H
686

Cod's
Head
Urhin
R571
R575
Slieve Miskish Mts
R572
Currygl

Allihies
Knockgour
481
R572
Castletownbere
Bere Haven
Ballynakilla
Lo

Dursey
Island
Cable
Car
Garinish
Ballydonegan Bay
Ballydonegan
Fair
Head
Bere
Island
Rerrin

Cahermore
R571

Dursey
Head
Crow
Head
Black
Ball Head
Ballyroo

Muntervary or
Sheeps Head

**4**
Film Location:
STAR WARS EPISODE VIII:
THE LAST JEDI
Three Castle Head
208
K

MIZEN HEAD
SIGNAL STATION
Mizen
Head
Barley Cove

0    2    4    6 miles
0  2  4  6  8  10 km

**A**   **B**   **C**

Knocknaralna
794
668
Ballinamult
Seefin 728
Furraleigh
Newtown
WATERF
(PORT LÁIR
D
37
Toraneena
Kilbrien
E
Kilmacthomas
38
Fa
agh
F
llymorris
TRAMORE
HO. GDNS.
R685
Clohern
204
Kilrossanty
Lemybrien
Kill
Dunhill
FENOR
BOG
R672
Millstreet
Ballylaneen
DUNHILL
CASTLE
Fennor
TRAMORE
R669
R671
DEISE GREENWAY
R677
METAL MAN
TOWER
Tramore (Trá Mhór)
CAPPOQUIN
HOUSE &
GARDEN
Cappoquin
Modelligo
R675
Annestown
Ballymacaw
Dur
N72
Bunmahon
Lismore
Kilgobnet
Stradbally
Dunabrattin
Bay
Great
Newtown
Head
Brownstown
Brownstown
Head
HERITAGE
CENTRE
Ballymacmague
Ballyvoyle
Film Location:
REDWATER
Dunabrattin
Head
COPPER
COAST
GEOPARK
bridge
Whitechurch
Dungarvan
(Dún Garbhán)
Garrynageragh
Ballyvoyle
Head
R671
MUS.
Ballynacourty
Clonea Bay
Villierstown
Dunmoon
Aglish
Drum Hills
302
R674
Dungarvan Harbour
Helvick
Head
An Goirtín
(Gorteen)
An Rinn
(Ringville)
he Pike
Knockanore
Muggort's Bay
Boola
Cross
N25
Loiscreán
(Loskeran)
Clashmore
Ré na gCloichín
(Reanaclogheen)
Mine Head
Grange
R673
HERITAGE
CEN
Curragh
Ardmore
Bay
E CHURCH
ST MARY
Kinsalebeg
Ardmore
agh
45 (72)
Youghal
(Eochaill)
Ardmore Head
ROUND TOWER
& CATHEDRAL
ortaroo
Film Location:
MOBY DICK
Whiting
Bay
Ballymadog
Youghal Bay
Ballymacoda
Knockadoon
Head
credan
61
oe
Ballycotton
Bay
ry
tton

1

2

3

4

0        2        4        6 miles
0    2    4    6    8    10 km

D                    E                    F

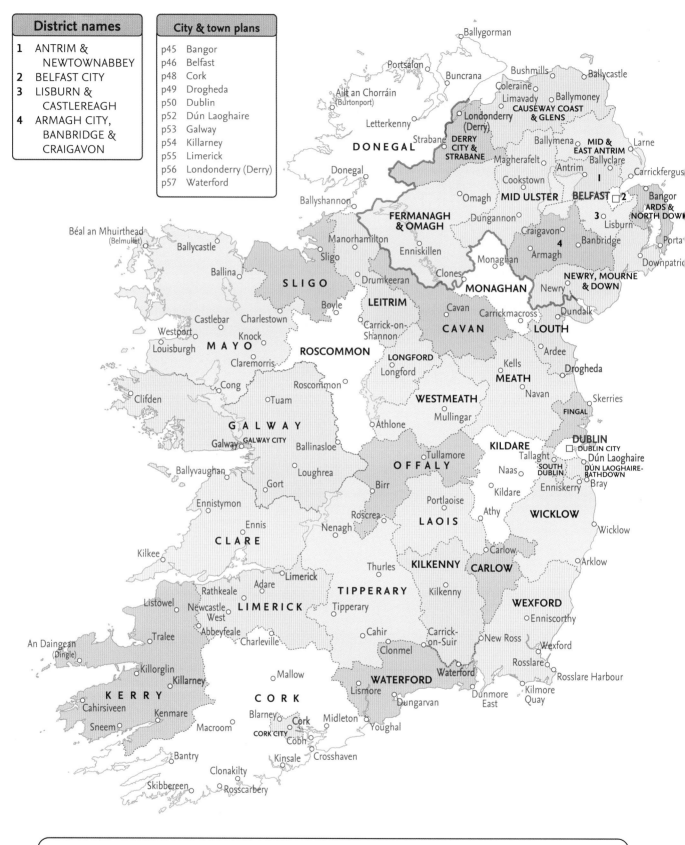

| District names | City & town plans |
|---|---|
| 1  ANTRIM & NEWTOWNABBEY | p45  Bangor |
| 2  BELFAST CITY | p46  Belfast |
| 3  LISBURN & CASTLEREAGH | p48  Cork |
| 4  ARMAGH CITY, BANBRIDGE & CRAIGAVON | p49  Drogheda |
| | p50  Dublin |
| | p52  Dún Laoghaire |
| | p53  Galway |
| | p54  Killarney |
| | p55  Limerick |
| | p56  Londonderry (Derry) |
| | p57  Waterford |

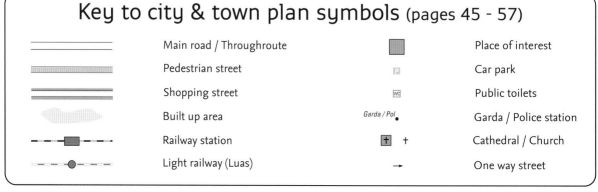

## Key to city & town plan symbols (pages 45 - 57)

| | | | |
|---|---|---|---|
| Main road / Throughroute | | Place of interest |
| Pedestrian street | | Car park |
| Shopping street | | Public toilets |
| Built up area | Garda / Pol | Garda / Police station |
| Railway station | | Cathedral / Church |
| Light railway (Luas) | → | One way street |

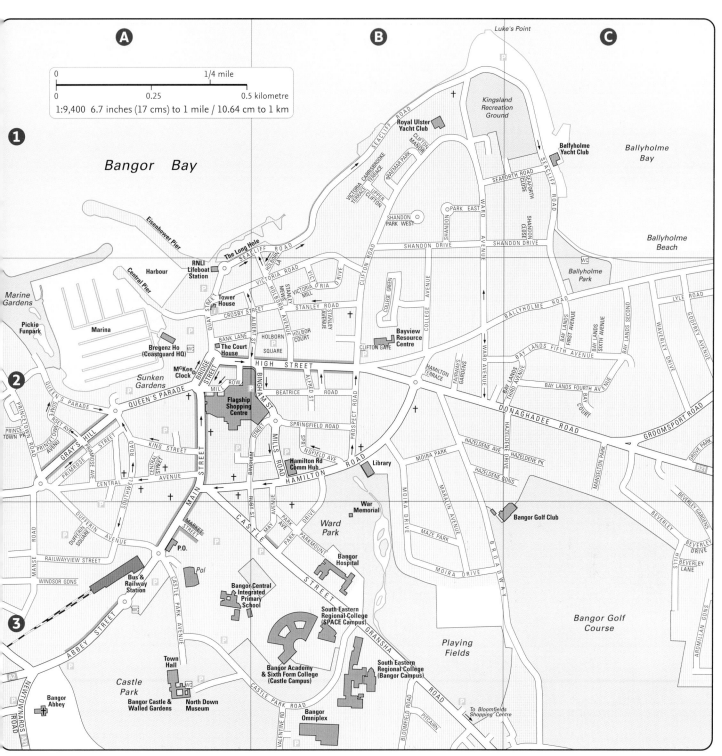

Bangor (population 61,011) is sited on the southern shore of Belfast Lough, 12 miles (19 km) east of Belfast and has historically been one of Northern Ireland's most popular holiday destinations. It is the home to two yacht clubs and has Northern Ireland's largest marina which hosts a variety of major sailing and tourist events each year.

## INDEX TO STREET NAMES

| | | | | | | | |
|---|---|---|---|---|---|---|---|
| Abbey St | A3 | Bingham St | B2 | Godfrey Av | C2 | May Av | B3 | Seaforth Close | C1 |
| Albert St | B2 | Bloomfield Rd | B3 | Gransha Rd | B3 | Maze Park | B3 | Shandon Close | C1 |
| Alfred St | B2 | Braemar Park | B1 | Gray's Hill | A2 | Mill Row | A2 | Shandon Dr | B1 |
| Ardmillan Gdns | C3 | Bridge St | A2 | Groomsport Rd | C2 | Mills Rd | B2 | Shandon Park |
| Ballyholme Rd | C2 | Broadway | B3 | Grove Park | C2 | Moira Dr | B2 | East | B1 |
| Bank La | A2 | Carrisbrooke Ter | B1 | Hamilton Rd | B2 | Moira Park | B2 | Shandon Park |
| Bay Lands Fifth Av | C2 | Castle Park Av | A3 | Hamilton Ter | B2 | Newtownards Rd | A3 | West | B1 |
| Bay Lands First Av | C2 | Castle Park Rd | B3 | Hazeldene Ave | B2 | Park Ave | B3 | Somerset Ave | A2 |
| Bay Lands | | Castle St | A3 | Hazeldene Dr | C2 | Park Dr | B3 | Southwell Rd | A2 |
| Fourth Av | C2 | Central Av | A2 | Hazeldene Gdns | B2 | Parkmount | B3 | Springfield Ave | B2 |
| Bay Lands | | Central St | A2 | Hazeldene Pk | C2 | Pitcairn Ave | B3 | Springfield Rd | B2 |
| Second Av | C2 | Clifton Gate | B2 | High St | A2 | Primrose Ave | A2 | Stanley Av | B2 |
| Bay Lands Sixth Av | C2 | Clifton Manor | B1 | Holborn Av | B2 | Primrose St | A2 | Stanley Mews | B2 |
| Bay Lands Third Av | C2 | Clifton Rd | B2 | Holborn Court | B2 | Princetown Park | A2 | Stanley Rd | B2 |
| Bay Lands Court | B2 | College Av | B2 | Holborn Lane | B2 | Princetown Rd | A2 | Upper Clifton | B1 |
| Beatrice Rd | B2 | College Green | B2 | Holborn Square | B2 | Princetown Av | A2 | Valentine Rd | B3 |
| Beverley Dr | C3 | Crosby St | A2 | King St | A2 | Prospect Rd | B2 | Victoria Dr | B2 |
| Beverley Gdns | C2 | Donaghadee Rd | B2 | Lyle Rd | C2 | Quay St | A2 | Victoria Mill | B2 |
| Beverley Hills | C3 | Dufferin Av | A3 | Main St | A2 | Queen's Parade | A2 | Victoria Ter | B1 |
| Beverley La | C3 | Dufferin Square | A3 | Manse Rd | A3 | Railwayview St | A3 | Victoria Rd | B2 |
| Bingham St | B2 | Fairways Gdns | B2 | Manselton Park | C2 | Ruby St | B2 | Ward Av | B1 |
| | | | | Maralin Av | B2 | Seacliff Rd | B1 | Waverley Dr | C2 |
| | | | | Market St | A3 | Seaforth Rd | B1 | Windsor Gdns | A3 |

## INDEX TO PLACES OF INTEREST

| | | | |
|---|---|---|---|
| Ballyholme Yacht | | RNLI Lifeboat | |
| Club | C1 | Station | A2 |
| Bangor Abbey | A3 | Royal Ulster | |
| Bangor Academy | | Yacht Club | B1 |
| & Sixth Form | | South Eastern | |
| College | | Regional College | |
| (Castle Campus) | B3 | (Bangor Campus) | B3 |
| Bangor Castle & | | South Eastern | |
| Walled Gardens | A3 | Regional College | |
| Bangor Golf Club | C3 | (SPACE Campus) | B3 |
| Bangor Hospital | B3 | The Court House | A2 |
| Bangor Omniplex | B3 | The Long Hole | A1 |
| Bregenz House | | Tower House | A2 |
| (Coastguard HQ) | A2 | Town Hall | A3 |
| Castle Park | A3 | War Memorial | B3 |
| Central Pier | A2 | Ward Park | B3 |
| Eisenhower Pier | A1 | | |
| Flagship Shopping | | | |
| Centre | A2 | | |
| Hamilton Road | | | |
| Community Hub | B2 | | |
| Library | B2 | | |
| Marine Gardens | A2 | | |
| McKee Clock | A2 | | |
| North Down | | | |
| Museum | A3 | | |
| Pickie Funpark | A2 | | |

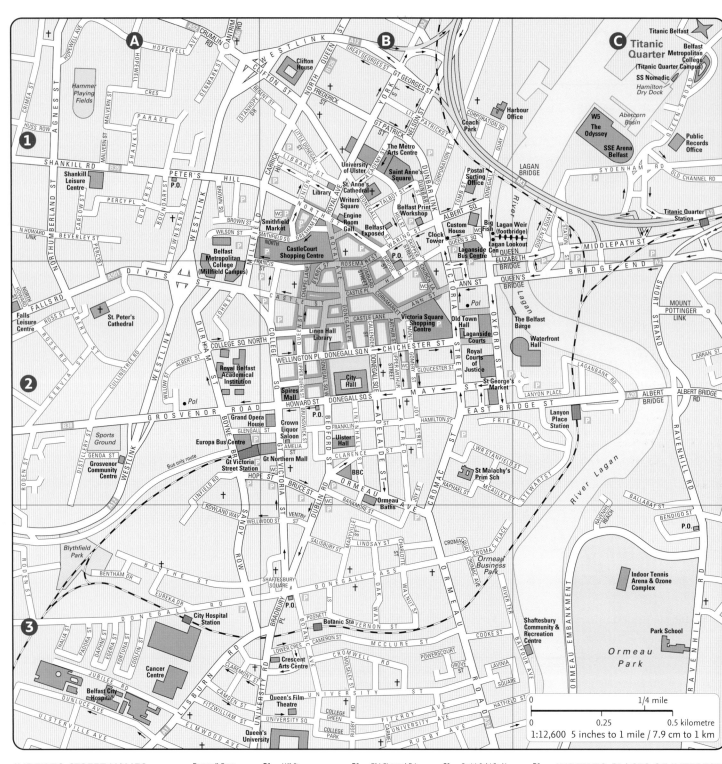

## INDEX TO STREET NAMES

Academy St B1
Adelaide St B2
Agnes St A1
Albert Bridge C2
Albert Bridge Rd C2
Albert Sq B1
Albert St A2
Alfred St B2
Amelia St B2
Ann St B2
Antrim Rd A1
Arran St C2
Arthur St B2
Balfour Av C3
Ballarat St C2
Bankmore St B2
Bank St B1
Bedford St B2
Bendigo St C3
Bentham Dr A3
Berry St B2
Beverley St A1
Blythe St A3
Botanic Av B3
Boundary St A1
Boyne Bridge A2
Bradbury Pl B3
Bridge End C2
Bridge St B1
Brown Sq A1
Brown St A1
Bruce St B2
Brunswick St B2
Callender St B2
Camden St A3
Cameron St B3

Carlow St A1
Carmel St B3
Carrick Hill B1
Castle Pl B2
Castle St B2
Chapel La B2
Charlotte St B2
Chichester St B2
Church La B2
Claremont St A3
Clarence St B2
Clifton St B1
College Grn B3
College Pk B3
College Sq B2
College Sq N A2
Cooke St A3
Coolfin St A3
Cornmarket B2
Corporation Sq B1
Corporation St B1
Crimea St A2
Cromac Av B3
Cromac Pl B3
Cromac Quay B3
Cromac St B2
Crumlin Rd A1
Cullender St B2
Cullingtree Rd A2
Daphne St A3
Denmark St B1
Distillery St A2
Divis St A2
Donegall Pass B3
Donegall Pl B2

Donegall Quay B1
Donegall Rd A3
Donegall Sq E B2
Donegall Sq N B2
Donegall Sq S B2
Donegall Sq W B2
Donegall St B1
Dover St A1
Dublin Rd B3
Dunbar Link B1
Dunbar St B1
Dunluce Av A3
Durham St A2
East Bridge St B2
Egeria St A3
Elmwood Av B3
Eureka Dr A3
Exchange St B1
Falls Rd A2
Fitzroy Av B3
Fitzwilliam St B3
Fortuna St A3
Francis St B1
Franklin St B2
Frederick St B1
Friendly St B3
Genoa St A2
Glengall St A2
Gloucester St B2
Great Georges St B1
Great Patrick St B1
Great Victoria St A2
Gresham St B1
Grosvenor Rd A2
Grove St B3
Hamilton St B2
Hatfield St B3
High St B2

Hill St B1
Hope St A2
Hopewell Av A1
Hopewell Cres A1
Howard St B2
Huss Row A1
John St A2
Joy St B2
Jubilee Rd A3
Laganbank Rd C2
Lanyon Place C2
Lavinia Sq B3
Library St B1
Lindsay St B3
Linenhall St B2
Linfield St A2
Lisburn Rd A3
Little Donegall St B1
Little Patrick St B1
Lombard St B2
Lower Cres B3
Lower Stanfield St B2
Malvern St A1
Maryville St B3
May St B2
McAuley St B2
McClure St B3
Middlepath St C1
Millfield A1
Montgomery St B2
Mount Pottinger Link C2
Nelson St B1
North Howard Link A1
North Howard St A1
North Queen St B1
North St B1
Northumberland St A1
Oak Way B3

Old Channel Rd C1
Ormeau Av B2
Ormeau Embankment C3
Ormeau Rd B3
Oxford St B2
Padora St A3
Percy Pl A1
Percy St A1
Peter's Hill A1
Posnett St B3
Powerscourt B3
Queen Elizabeth Bridge B1
Queen St B2
Queen's Bridge C2
Queen's Quay C1
Queen's Sq B1
Raphael St B3
Ravenhill Rd B3
Ravenhill Reach C3
Regent St A1
River Ter B3
Roden St A3
Rosemary St B2
Ross Rd A2
Ross St A2
Rowland Way A3
Royal Av B1
Rugby Av B3
Rugby Rd B3
Salisbury St A3
Sandy Row A2
Servia St A2
Shaftesbury Sq A3
Shankill Par A1
Shankill Rd A1
Short Strand C2
Skipper St B1

Smithfield Sq N B1
Stanhope St A1
Station St C1
Stewart St C2
Sydenham Rd C1
Talbot St B1
Thalia St A3
Tomb St B1
Townsend St A1
Ulsterville Av A3
Union St B1
University Av B3
University Rd B3
University Sq B3
Upper Arthur St B2
Upper Queen St B2
Ventry St B3
Vernon St B3
Victoria St B2
Walnut St B3
Waring St B1
Wellington Pl B2
Wellwood St A3
Westlink B1
West St B1
Willow St A2
Wilson St A1
Wolseley St A3
York St B1

## INDEX TO PLACES OF INTEREST

B.B.C. B2
Belfast City Hospital A3
Belfast Exposed B1
Belfast Metropolitan
   College (Millfield
   Campus) A1
Belfast Metropolitan
   College (Titanic
   Quarter Campus) C1
Belfast Print
   Workshop B1
Big Fish B1
Botanic Station B3
Cancer Centre A3
CastleCourt
   Shopping Centre B1
Cinema B2
City Hall B2
City Hospital Station A3
Clock Tower B1
College of Further
   Education C3
Crescent Arts Centre B3
Custom House B1
Engine Room Gallery B1
Europa Bus Centre A2
Falls Leisure Centre A2
General Post Office A2
Grand Opera House A2
Great Northern Mall A2
Great Victoria St
   Station A2
Grosvenor Community
   Centre A2
Indoor Tennis Arena &
   Ozone Complex C3

Lagan Weir B1
Lagan Lookout Centre B1
Laganside Courts B2
Lanyon Place Station C2
Linen Hall Library B2
Millfield Technical
   College A2
Ormeau Baths B2
Ormeau Park C3
Ozone C3
Public Records Office C1
Queen's University B3
Royal Belfast Academical
   Institution A2
Royal Courts of Justice B2
St. Anne's Cathedral B1
St. Anne's Square B1
St. George's Market B2
St. Peter's Cathedral A2
Shaftesbury Community
   & Recreation Centre C3
Shankill Leisure Centre A1
Smithfield Market B1
Spires Mall B2
SSE Arena Belfast/
   The Odyssey C1
The Belfast Barge B2
The Metro Arts Centre B1
Titanic Belfast C1
Titanic Quarter Station C1
Ulster Hall B2
University of Ulster B1
Victoria Square
   Shopping Centre B2
W5 C1
Waterfront Hall C2

## General information

Population 343,542 (2019). The capital of Northern Ireland. Sited on the River Lagan at the head of Belfast Lough, the city grew from a small village during the industrial revolution with industries such as linen, rope making and shipbuilding.

## Tourist information

The **Visit Belfast Welcome Centre** at 9 Donegall Square North, Belfast BT1 5GB ☎ 028 9024 6609 www.visitbelfast.com provides an information and accommodation booking service.

The Northern Ireland Tourist Board website is available at www.discovernorthernireland.com

## Getting around

Two bus services run in and around Belfast city. Ulsterbus transports people in and out of the city and serves all major towns and villages. Metro runs around the city, departing and terminating in the city centre. For information on Metro, Ulsterbus and Northern Ireland Railways contact **Translink** ☎ 028 9066 6630 www.translink.co.uk

The main railway station in Belfast is Central Station, East Bridge Street. For rail enquiries contact **Translink** ☎ 028 9066 6630 www.translink.co.uk

Belfast City Hall

Photo © Northern Ireland Tourist Board

## Places of interest

The places of interest section (pages 6-12) includes **Belfast Cathedral** but there are of course many more places to visit within Belfast city centre. Buildings of architectural interest include **Belfast City Hall** which dominates Donegall Square. It is a striking classical Renaissance style building completed in 1906 and its great copper dome is a landmark throughout the city. Nearby, the **Linen Hall Library** is Belfast's oldest library and is the leading centre for Irish and local studies in Northern Ireland. Specialising in Irish culture and politics, it also has a unique collection of early printed books from Belfast and Ulster. At Queen's University the **Queen's Welcome Centre** provides information about the university and presents a varied programme of exhibitions. Located at the heart of the campus in the Lanyon Room, the centre is named after Charles Lanyon who was the architect of the main Queen's building and many other public buildings in Ireland.

The extravagant **Crown Liquor Saloon** on Great Victoria Street is of historic interest. It dates from Victorian times and is one of the most famous public houses in Belfast. More recent attractions include the **SSE Arena - The Odyssey**, a multi-functional entertainment centre which includes an 10,800 seat arena, the **W5** interactive discovery centre, cinema and leisure complex. Another well known city centre attraction are the colourful hand painted political murals of West Belfast. They adorn the walls and gable ends of many houses expressing the political viewpoints of the Protestant Shankill Road and the Catholic Falls Road and have become just as much a part of the tourist industry as the more traditional sites of the city. Black cab tours, with commentaries and photo stops, are available to view them.

To the north is **Belfast Castle** which overlooks the city from 122m (400ft) above sea level. Completed in 1870 by the 3rd Marquis of Donegall, this magnificent sandstone castle was refurbished over a 10 year period by Belfast City Council at a cost of more than £2m and was reopened to the public in 1988. East of the city is **Stormont Parliament Buildings**, the home of the Northern Ireland Parliament. The Main Hall is open to the public and tours for groups can be arranged in advance.

**Titanic Belfast** is an iconic building in the heart of Belfast's docklands, which provides visitors with a unique insight into all things Titanic. It tells the story of Titanic from her construction in Belfast to her fateful maiden voyage using stunning special effects, innovative interactive features and full-scale reconstructions.

## Theatres, concert halls & festivals

The Belfast **Waterfront Hall** ☎ 028 9033 4455 is Northern Ireland's premier concert and conference centre which covers a wide variety of entertainment in its flagship building. The **Grand Opera House** on Great Victoria Street ☎ 028 9024 1919 stages opera, drama, musicals, concerts and pantomime. The declining opulence of the building was restored in 1980 to transform it into a modern theatre whilst still retaining its lavish Victorian interior. **Ulster Hall**, first built in 1862 on Bedford Street ☎ 028 9033 4455, with its interior dominated by a massive English theatre organ, has been a favourite venue for concerts for over 140 years. The **King's Hall** ☎ 028 9066 5225 is another venue for concerts and gala shows. Many international and national events such as the Belfast Telegraph Ideal Home Exhibition and the Ulster Motor Show are held there. The main city centre theatre is the **Lyric Theatre** ☎ 028 9038 1081 which presents both classic and contemporary plays with an emphasis on Irish productions.

Annual events and festivals held in Belfast include the **Belfast International Arts Festival** ☎ 028 9089 2261 which is held in late autumn at the campus of Queen's University and other city venues. Hosting international theatre, dance, music and comedy, it is Ireland's largest arts festival. The **Cathedral Quarter Arts Festival** ☎ 028 9023 2403 is held in May to celebrate the best of the local talent as well as new international work.

## Shopping

The main city centre shopping area is Donegall Place, most of which is pedestrianised. Belfast is also renowned for its selection of malls and shopping centres. **CastleCourt** in Royal Avenue is Northern Ireland's largest shopping centre, with over 70 shops extending over 3.4ha (8.5 acres). Opposite CastleCourt is the modern **Smithfield Market** which replaced the old Victorian market destroyed by fire in 1974. The **Spires Centre and Mall** was refurbished in 1992 to become one of Belfast's most attractive buildings and is the place to shop for designer fashion and giftware. Belfast's newest shopping centre is **Victoria Square Shopping Centre**, located in the city's Southern Quarter.

## Parks & gardens

**Ormeau Park** opened in 1871 and is the largest park in the centre of the city. South of the city the **National Botanic Gardens** are one of Belfast's most popular parks. The restored Palm House was built in 1840 and is one of the earliest examples of a curved glass and wrought iron glasshouse. North of the city **Belfast Zoo** is set in landscaped parkland on the slopes of Cave Hill. Over 120 species are housed there and the zoo increasingly focuses on wildlife facing extinction so has specialised collections with breeding programmes for endangered species.

## Telephoning

If telephoning from Great Britain or Northern Ireland use the area code and telephone number. If telephoning from the Republic of Ireland replace 028 with 048 and follow with the required telephone number.

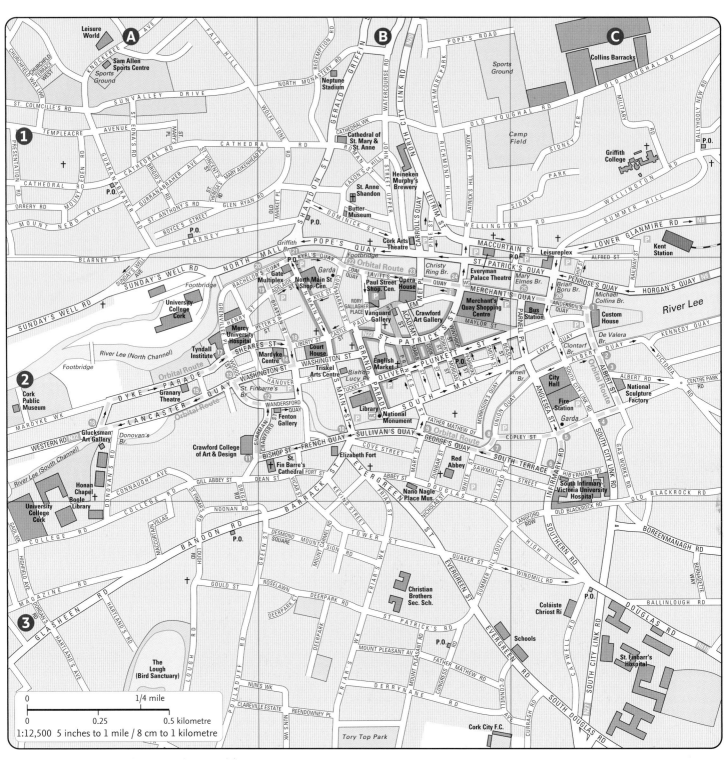

Cork (population 208,669) is Ireland's third largest city. The Irish form is Corcaigh, meaning marshy place, though the marshes are no longer in evidence. The city developed on the estuary of the River Lee and is now the cultural capital of the south west as well as a major commercial centre.

## INDEX TO STREET NAMES

| | | | | | | | | | | | |
|---|---|---|---|---|---|---|---|---|---|---|---|
| Abbey St | B2 | Churchfield Ter W | A1 | Evergreen Rd | B3 | High St | C3 | Mount Carmel Rd | B3 | Princes St | B2 | South City Link Rd | C2 |
| Academy St | B2 | Churchfield Way | | Evergreen St | B2 | Highfield Av | A3 | Mount Eden Rd | A1 | Quaker St | B3 | South Douglas Rd | C3 |
| Adelaide St | B2 | Lwr | A1 | Fair Hill | A1 | Horgan's Quay | C2 | Mount Nebo Av | A1 | Railway St | B1 | South Main St | B2 |
| Albert Quay | C2 | Clareville Estate | A3 | Father Mathew | | Infirmary Rd | C2 | Mount Pleasant Ave | B3 | Rathmore Pk | B1 | South Mall | B2 |
| Albert Rd | C2 | Coal Quay | B2 | Quay | B2 | John Street Upper | B1 | Mount Pleasant Rd | B3 | Reendowney Pl | A3 | South Terr | B2 |
| Albert St | C2 | College Rd | A3 | Father Mathew Rd | B3 | Kennedy Quay | C2 | Mount Sion Rd | B3 | Richmond Hill | B1 | Southern Rd | C3 |
| Alfred St | C1 | Congress Rd | B3 | Fort St | B2 | Kevins St | B2 | Nicholas St | B2 | Robert St | B1 | Sullivan's Quay | B2 |
| Anderson's Quay | C2 | Connaught Av | A2 | French Quay | B2 | Knockfree Av | A1 | Noonan Rd | A3 | Roman St | B1 | Summer Hill | C1 |
| Anglesea St | B2 | Cook St | B2 | Friar's Wk | B3 | Kyle St | B2 | North Main St | B2 | Rory Gallagher Pl | B2 | Summer Hill S | B3 |
| Audley Pl | B1 | Copley St | C2 | Friar St | B2 | Kyrl's Quay | B1 | North Mall | A2 | Roselawn | A3 | Sunday's Well Av | A2 |
| Bachelor's Quay | A2 | Cornmarket St | B2 | Gaol Wk | A3 | Lancaster Quay | A2 | North Monastery | | Rutland St | B2 | Sunday's Well Rd | A2 |
| Ballinlough Rd | C3 | Cove St | B2 | Gas Works Rd | B1 | Langford Row | C3 | Rd | B1 | St. Anthony's Rd | A1 | Templeacre Ave | A1 |
| Ballyhooly New Rd | C1 | Curragh Rd | C3 | George's Quay | B2 | Lapp's Quay | C2 | North City Link Rd | B1 | St. Colmcille's Rd | A1 | Tower St | B3 |
| Bandon Rd | A3 | Dean St | A2 | Gerald Griffin St | B1 | Lavitt's Quay | B2 | North Main St | B2 | St. Edna's Rd | A3 | Tuckey St | B2 |
| Barrack St | B3 | Deerpark | B3 | Gill Abbey St | A2 | Leitrim St | B1 | Nun's Wk | A3 | St. Finbar's Rd | A3 | Union Quay | B2 |
| Beasly St | B2 | Deerpark Ct | B3 | Glasheen Rd | A3 | Liberty St | B2 | O'Connell Av | B3 | St. Mary's Pl | B1 | Valley Dr | A1 |
| Bernadette Way | C3 | Deerpark Rd | B3 | Glen Ryan Rd | A1 | Lough Rd | A3 | Old Blackrock Rd | C3 | St. Mary's Rd | B1 | Vicar St | B2 |
| Bishop St | B3 | Derrynane Rd | B3 | Glenville Rd | A2 | Lower Glanmire Rd | C1 | Old Market Pl | B1 | St. Patrick's Quay | B2 | Victoria Rd | C2 |
| Blarney St | A1 | Desmond Sq | B3 | Gould St | A3 | Mac Curtain St | B2 | Oliver Plunkett St | B2 | St. Patrick's Rd | B3 | Wandersford Quay | B2 |
| Boreenmanagh Rd | C3 | Dominick St | B1 | Grand Parade | B2 | MacCurtain Villas | A3 | Orrery Rd | A1 | St. Patrick's St | B2 | Washington St | A2 |
| Boyce's St | A2 | Donovan's Rd | A2 | Grattan St | B2 | Magazine Rd | A3 | Parnell Pl | C2 | St. Theresa's Rd | A3 | Washington St W | A2 |
| Brigid's St | B1 | Dorgan's Rd | A3 | Green St | B3 | Mardyke Wk | A2 | Patrick's Hill | B1 | St. Vincent's St | A2 | Watercourse Rd | B1 |
| Capwell Rd | C3 | Douglas Rd | C3 | Gregg Rd | A2 | Marlborough St | B2 | Paul St | B2 | Sawmill St | B2 | Wellington Rd | B1 |
| Carroll's Quay | B2 | Douglas St | B2 | Gurrana Braher Av | A1 | Mary Aikenhead Pl | A2 | Pembroke St | B2 | Shandon St | B1 | Western Rd | A2 |
| Castle St | B2 | Dunbar St | B2 | Gurrana Braher Rd | A1 | Mary St | B2 | Penrose's Quay | C2 | Sharman Crawford | | White St | B2 |
| Cathedral Rd | B1 | Dyke Par | A2 | Hanover St | B2 | Maylor St | B2 | Peter's St | B2 | St | B2 | Windmill Rd | C3 |
| Cathedral Wk | B1 | Eason's Hill | B1 | Hartland's Av | A3 | Merchant's Quay | B2 | Pine St | B1 | Sheares St | A2 | Winthrop St | B2 |
| Centre Park Rd | C2 | Emmet Pl | B2 | Hartland's Rd | A3 | Military Rd | C1 | Pope's Quay | B1 | Ship St | B2 | Wolfe Tone Rd | B1 |
| | | | | Henry St | B2 | Morgan St | B2 | Pope's Rd | B1 | Sidney Pk | C1 | Woods St | A2 |
| | | | | Hibernian Rd | C2 | Morrison's Quay | B2 | Pouladuff Rd | A3 | Sidney Ter | C1 | Youghal Old Rd | B1 |
| | | | | | | | | Presentation Rd | A1 | Smith St | B2 | | |

## INDEX TO PLACES OF INTEREST

| | | | |
|---|---|---|---|
| Bon Secours Hospital | A3 | Mercy University | |
| Boole Library | A2 | Hospital | A2 |
| Bus Station | C2 | Nano Nagle | |
| Butter Museum | B1 | Place Museum | B2 |
| Camp Field | C1 | National Monument | B2 |
| Cathedral of St. Mary & | | National Sculpture | |
| St. Anne | B1 | Factory | C2 |
| City Hall | C2 | Neptune Stadium | B1 |
| Collins Barracks | C1 | North Main Street | |
| Cork Arts Theatre | B1 | Shopping Centre | B2 |
| Cork Public Museum | A2 | Opera House | B2 |
| Court House | B2 | Paul Street Shopping | |
| Crawford College of | | Centre | B2 |
| Art and Design | A2 | Red Abbey | B2 |
| Crawford Art Gallery | B2 | St. Anne Shandon | B1 |
| Custom House | C2 | St. Fin Barre's | |
| Elizabeth Fort | B2 | Cathedral | B2 |
| English Market | B2 | St. Finbarr's Hospital | C3 |
| Everyman Palace | | Sam Allen Sports | |
| Theatre | B1 | Centre | A1 |
| Fenton Gallery | B2 | South Infirmary-Victoria | |
| Gate Multiplex | B2 | University Hospital | C2 |
| General Post Office | B2 | The Lough | |
| Glucksman | | (Bird Sanctuary) | A3 |
| Art Gallery | A2 | Tory Top Park | B3 |
| Granary Theatre | A2 | Trishel Arts Centre | B2 |
| Heineken Murphy's | | Tyndall Institute | A2 |
| Brewery | B1 | University College, | |
| Honan Chapel | A2 | Cork | A2 |
| Kent Station | C1 | | |
| Leisureplex | C2 | | |
| Leisure World | A1 | | |
| Mardyke Centre | B2 | | |
| Merchant's Quay | | | |
| Shopping Centre | B2 | | |

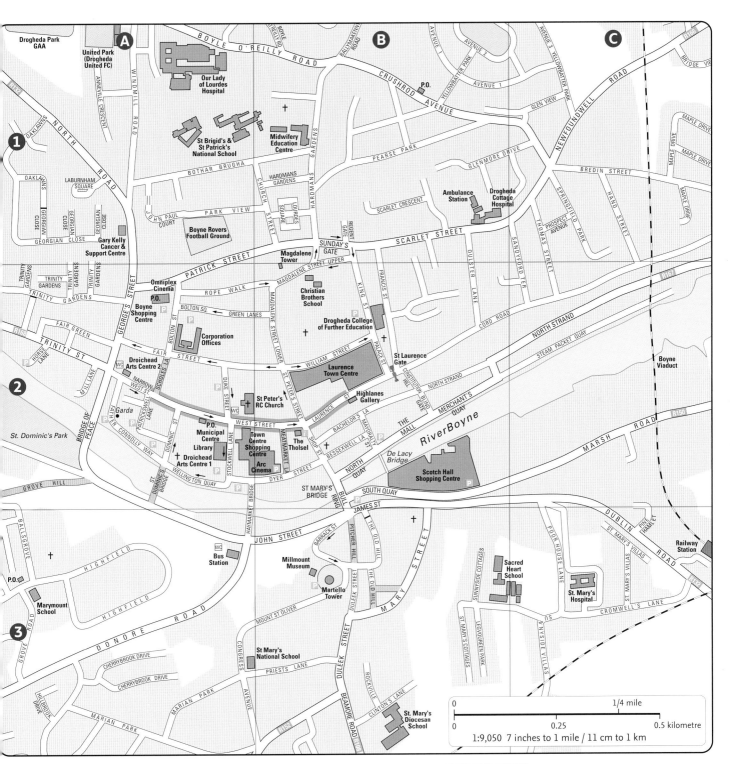

Drogheda (population 40,956) developed as a seaport and centre of strategic importance at the mouth of the River Boyne, 32 miles (51km) north of Dublin. It has gained increased commercial importance by being on the significantly improved Rosslare-Dublin-Dundalk-Belfast route which has brought Drogheda within easy commuting range of Dublin.

## INDEX TO STREET NAMES

| | | | | | | | | | |
|---|---|---|---|---|---|---|---|---|---|
| Anneville Cres | A1 | Clinton's La | B3 | Green Lanes | A2 | Marsh Rd | C2 | Prospect Av | C1 | Sunnyside Villas | C3 |
| Avenue 1 | B1 | Congress Av | A3 | Grove Hill | A2 | Mary St | B3 | Regent Gate | B1 | The Old Hill | B3 |
| Avenue 2 | B1 | Constitution Hill | B2 | Grove Rd | A3 | Mayoralty St | B2 | Rockville | B3 | The Mall | B2 |
| Avenue 3 | C1 | Cord Rd | B2 | Hand St | C1 | Meatmarket La | B2 | Rope Walk | A2 | Thomas St | C1 |
| Bachelor's La | B2 | Cromwell's La | C3 | Hardmans Gdns | B1 | Merchant's Quay | B2 | St. Dominic's | | Trinity Gdns | A2 |
| Ballsgrove | A3 | Crushrod Av | B1 | Haymarket Bridge | A3 | Mill La | A2 | Bridge | A2 | Trinity St | A2 |
| Ballymakenny Rd | B1 | Dominic St | A2 | Highfield | A3 | Mount St Oliver | B3 | St. Mary's Bridge | B2 | Wellington Quay | A2 |
| Barrack St | B3 | Donore Rd | A3 | Hillbrook Dr | A3 | Narrow West St | A2 | St. Mary's Cottages | B3 | West St | B2 |
| Beamore Rd | B3 | Duke St | A2 | Horse La | A2 | Newfoundwell Rd | C1 | St. Mary's Villas | C3 | William St | B2 |
| Bessexwell La | B2 | Duleek St | B3 | James St | B3 | North Rd | A1 | St. Peter's St | B2 | Windmill Rd | A1 |
| Blind Gate | B2 | Dyer St | B2 | John Paul Ct | A1 | North Strand | C2 | Sandyford Ter | C1 | Yellowbatter Pk | B1 |
| Bolton Sq | A2 | Fair Green | A2 | John St | B3 | North Quay | B2 | Scarlet Cres | B1 | Yellowbatter Pk | C1 |
| Bolton St | A2 | Fair St | A2 | King St | B2 | Oaklawns | A1 | Scarlet St | B1 | |
| Bothar Brugha | A1 | Father Connolly | | Laburnham Sq | A1 | Oulster La | B1 | Scholes La | A2 | |
| Boyle O'Reilly Rd | A1 | Way | A2 | Laurence St | B2 | Palace St | B2 | Shop St | B2 | |
| Boyne O'Reilly Rd | A1 | Francis St | B2 | Legvoureen Pk | A3 | Park View | A1 | South Quay | B2 | |
| Bredin St | C1 | George's St | A2 | Lourdes Square | B1 | Patrick St | A2 | Springfield Pk | C1 | |
| Bridge Of Peace | A2 | Georgian Cl | A1 | Magdalene St | | Patrickswell La | A2 | Steam Packet | | |
| Bridge View | C1 | Glen View | C1 | Lower | B2 | Pearse Pk | B1 | Quay | C2 | |
| Bull Ring | B2 | Glenmore Dr | B1 | Magdalene St | | Pine Hamlets | C3 | Stockwell La | A2 | |
| Cherrybrook Dr | A3 | | | Upper | B1 | Pitcher Hill | B3 | Sunday's Gate | B1 | |
| Church St | B1 | | | Maple Dr | C1 | Poor House La | C3 | Sunnyside | | |
| | | | | Marian Pk | A3 | Priests La | B3 | Cottages | B3 | |

## INDEX TO PLACES OF INTEREST

| | | | |
|---|---|---|---|
| Arc Cinema | B2 | Municipal Centre | A2 |
| Boyne Rovers Football | | Omniplex Cinema | |
| Ground | A1 | (Boyne Shopping | |
| Boyne Shopping | | Centre) | A2 |
| Centre | A2 | Our Lady of | |
| Boyne Viaduct | C2 | Lourdes Hospital | A1 |
| Bus Station | A3 | Railway Station | C3 |
| Corporation | | Sacred Heart School | C3 |
| Offices | A2 | Scotch Hall | |
| Drogheda College | | Shopping Centre | B2 |
| of Further | | St Laurence Gate | B2 |
| Education | B2 | St Peter's RC Church | B2 |
| Drogheda Cottage | | St. Mary's Hospital | C3 |
| Hospital | B1 | The Tholsel | B2 |
| Drogheda Park GAA | A1 | Town Centre | |
| Droichead Arts | | Shopping Centre | A2 |
| Centre 1 | A2 | United Park | |
| Droichead Arts | | (Drogheda | |
| Centre 2 | A2 | United FC) | A1 |
| Gary Kelly Cancer & | | | |
| Support Centre | A1 | | |
| Highlanes Gallery | B2 | | |
| Laurence Town | | | |
| Centre | B2 | | |
| Magdalene Tower | B1 | | |
| Martello Tower | B3 | | |
| Midwifery Education | | | |
| Centre | B1 | | |
| Millmount Museum | B3 | | |

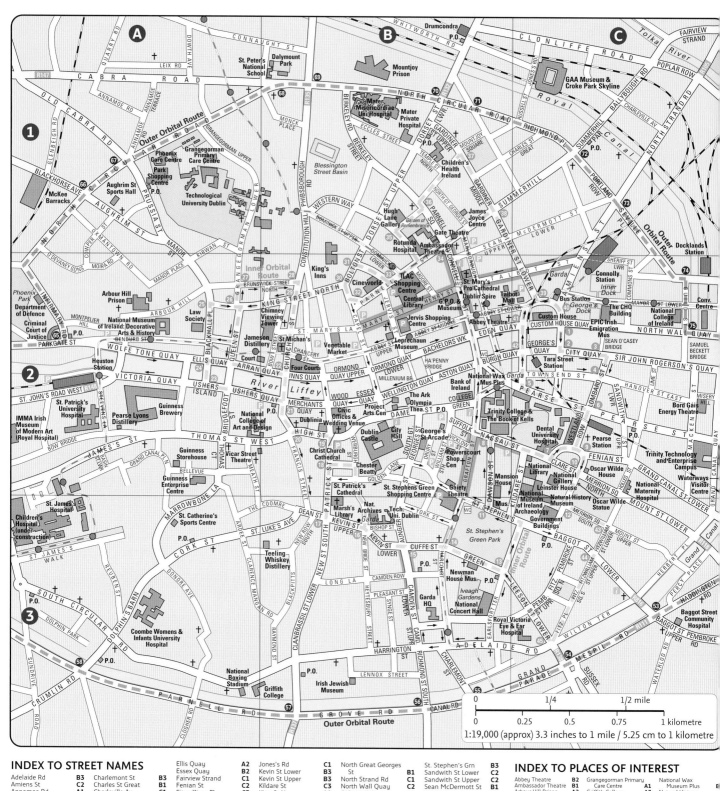

## INDEX TO STREET NAMES

| Street | Grid | | Street | Grid |
|---|---|---|---|---|
| Adelaide Rd | B3 | | Charlemont St | B3 |
| Amiens St | C2 | | Charles St Great | B1 |
| Annamoe Rd | A1 | | Charleville Av | C1 |
| Annamoe Ter | A1 | | Church St | B2 |
| Arbour Hill | A2 | | City Quay | C2 |
| Ardee St | A3 | | Clanbrassil St Lower | B3 |
| Arran Quay | A2 | | Clare St | C2 |
| Aston Quay | B2 | | Clarence Mangan Rd | A3 |
| Aughrim St | A1 | | Clogher Road | A3 |
| Aungier St | B3 | | Clonliffe Rd | B1 |
| Bachelors Wk | B2 | | College Grn | B2 |
| Baggot St Lower | C3 | | Commons St | C2 |
| Baggot St Upper | C3 | | Connaught St | A1 |
| Ballybough Rd | C1 | | Constitution Hill | B2 |
| Basin St Lower | A2 | | Cork St | A3 |
| Bellevue | A2 | | Cowper St | A1 |
| Benburb St | A2 | | Crumlin Rd | A3 |
| Berkeley Rd | B1 | | Cuffe St | B3 |
| Berkeley St | B1 | | Custom House Quay | C2 |
| Bishop St | B3 | | Dame St | B2 |
| Blackhall Pl | A2 | | Dawson St | B2 |
| Blackpitts | B3 | | Dean St | B3 |
| Blackhorse Av | A1 | | Dolphin Park | A3 |
| Bolton St | B2 | | Dolphin Rd | A3 |
| Bow Bridge | A2 | | Dolphin's Barn | A3 |
| Bow St | B2 | | Dominick St Lower | B1 |
| Bridgefoot St | A2 | | Dominick St Upper | B1 |
| Brunswick St N | A2 | | Donore Av | A3 |
| Burgh Quay | B2 | | Dorset St Lower | B1 |
| Cabra Rd | A1 | | Dorset St Upper | B1 |
| Camden Row | B3 | | Dowth Av | A1 |
| Camden St Lower | B3 | | Drury Street | B2 |
| Camden St Upper | B3 | | Earlsfort Ter | B3 |
| Canal Rd | B3 | | Eccles St | B1 |
| Capel St | B2 | | Eden Quay | B2 |

| Street | Grid | | Street | Grid |
|---|---|---|---|---|
| Ellis Quay | A2 | | Jones's Rd | C1 |
| Essex Quay | B2 | | Kevin St Lower | B3 |
| Fairview Strand | C1 | | Kevin St Upper | B3 |
| Fenian St | C2 | | Kildare St | C3 |
| Fitzwilliam Pl | C3 | | King St N | A2 |
| Fitzwilliam Sq N | C3 | | Kirwan St | A2 |
| Fitzwilliam Sq S | C3 | | Leeson St Lower | C3 |
| Fitzwilliam St Lower | C3 | | Leix Rd | A1 |
| Fitzwilliam St Upper | C3 | | Lennox St | B3 |
| Francis St | B2 | | Lime St | C2 |
| Gardiner St Lower | B1 | | Lombard St E | C2 |
| Gardiner St Middle | B1 | | Long La | B3 |
| Gardiner St Upper | B1 | | Macken St | C2 |
| George's Quay | C2 | | Manor Pl | A2 |
| Glenbeigh Rd | A1 | | Manor St | A1 |
| Golden La | B2 | | Marlborough St | B1 |
| Grafton St | B2 | | Marrowbone La | A3 |
| Grand Canal Pl N | A2 | | Mary St | B2 |
| Grand Canal Quay | C2 | | Mayor St Lower | C2 |
| Grand Canal St Lower | C2 | | Meath St | A2 |
| Grangegorman Lower | A2 | | Merchants Quay | B2 |
| Grangegorman Upper | A1 | | Merrion Sq N | C2 |
| Grove Rd | B3 | | Merrion Sq S | C3 |
| Guild St | C2 | | Mespil Rd | C3 |
| Haddington Rd | C3 | | Misery Hill | C2 |
| Hanover St E | C2 | | Moira Rd | A2 |
| Harcourt St | B3 | | Monck Pl | B1 |
| Harrington St | B3 | | Montpelier Hill | A2 |
| Henry St | B2 | | Moore St | B2 |
| Herbert Pl | C3 | | Mount St Lower | C3 |
| Heytesbury St | B3 | | Mount St Upper | C3 |
| High St | B2 | | Mountjoy Sq | B1 |
| Holles St | C2 | | Nassau St | B2 |
| Infirmary Rd | A2 | | New S St | B3 |
| Inns Quay | B2 | | New Bride St | B3 |
| James's St | A2 | | New Row S | B3 |
| Jervis St | B2 | | North Circular Rd | A1 |

| Street | Grid | | Street | Grid |
|---|---|---|---|---|
| North Great Georges St | C1 | | St. Stephen's Grn | B3 |
| North Strand Rd | C1 | | Sandwith St Lower | C2 |
| North Wall Quay | C2 | | Sandwith St Upper | C2 |
| O'Connell St | B2 | | Sean McDermott St | B1 |
| O'Devaney Gdns | A2 | | Seville Pl | C1 |
| Old Cabra Rd | A1 | | Sheriff St Lower | C2 |
| Ormond Quay Lower | B2 | | Sir John Rogerson's Quay | C2 |
| Ormond Quay Upper | B2 | | South Circular Rd | A3 |
| Oxmantown Rd | A2 | | South Great Georges St | B2 |
| Parkgate St | A2 | | Steevens La | A2 |
| Parnell Rd | A3 | | Suffolk St | B2 |
| Parnell Sq | B1 | | Summerhill | B1 |
| Parnell St | B1 | | Summerhill Par | C1 |
| Patrick St | B2 | | Sundrive Rd | A3 |
| Pearse St | C2 | | Synge St | B3 |
| Pembroke St | C3 | | Talbot St | B2 |
| Pembroke St Upr | C3 | | The Coombe | A3 |
| Percy Place | C3 | | Thomas Ct | A2 |
| Phibsborough Rd | B1 | | Thomas St W | A2 |
| Pleasant St | B3 | | Townsend St | C2 |
| Poplar Row | C1 | | Ushers Island | A2 |
| Portland Row | C1 | | Ushers Quay | A2 |
| Prussia St | A1 | | Victoria Quay | A2 |
| Quarry Rd | A1 | | Waterloo Rd | C3 |
| Raymond St | A3 | | Wellington Quay | B2 |
| Reuben St | A3 | | Werburgh St | B2 |
| Richmond Pl | B3 | | Western Way | B1 |
| Richmond St S | B3 | | Westland Row | C2 |
| Russell St | C1 | | Whitworth Rd | B1 |
| Rutland Avenue | A3 | | Wicklow Street | B2 |
| St. James's Wk | A2 | | William St South | B2 |
| St. John's Rd W | A2 | | Wilton Ter | C3 |
| St. Luke's Av | A3 | | Wolfe Tone Quay | A2 |
| St. Mary's La | B2 | | Wood Quay | B2 |
| | | | York St | B3 |

## INDEX TO PLACES OF INTEREST

| Place | Grid | | Place | Grid |
|---|---|---|---|---|
| Abbey Theatre | B2 | | Grangegorman Primary | |
| Ambassador Theatre | B1 | | Care Centre | A1 |
| Arbour Hill Prison | A2 | | Griffith College | A3 |
| Aughrim Street | | | Guinness Brewery | A2 |
| Sports Hall | A1 | | Guinness Enterprise | |
| Bank of Ireland | B2 | | Centre | A2 |
| Bord Gáis Energy | | | Guinness Storehouse | A2 |
| Theatre | C2 | | Hugh Lane Gallery | B1 |
| Book of Kells | B2 | | ILAC Shopping Centre | B2 |
| Bus Station | B2 | | IMMA Irish Museum | |
| Chester Beatty | B2 | | of Modern Art | A2 |
| Children's Health | | | James Joyce Centre | B1 |
| Ireland | B1 | | Jameson Distillery | B2 |
| Children's Hospital | A3 | | Jervis Shopping Centre | B2 |
| Chimney Viewing | | | King's Inns | B2 |
| Tower | B2 | | Leinster House | C2 |
| Christ Church | | | Leprechaun Museum | B2 |
| Cathedral | B2 | | Little Mus. of Dublin | B1 |
| Cineworld | B2 | | Mansion House | B2 |
| City Hall | B2 | | Mater Misericordiae | |
| Connolly Station | C2 | | University Hospital | B1 |
| Convention Centre | C2 | | Mater Private | |
| Coombe Womens & | | | Hospital | B1 |
| Infants University | | | McKee Barracks | A1 |
| Hospital | A3 | | Mountjoy Prison | B1 |
| Croke Park, GAA Mus & | | | National Archives | B3 |
| Croke Park Skyline | C1 | | National Boxing | |
| Custom House | C2 | | Stadium | A3 |
| Dalymount Park | B1 | | National Concert Hall | C3 |
| Docklands Station | C2 | | National Gallery | C2 |
| Dublin Castle | B2 | | National Library | C2 |
| Dublin Spire | B2 | | National Maternity | |
| EPIC Irish Emigration | | | Hospital | C2 |
| Museum | C2 | | National Museum of | |
| Four Courts | B2 | | Ireland: | |
| Gaiety Theatre | B2 | | Archaeology | C2 |
| Gate Theatre | B1 | | National Museum of | |
| General Post Office | | | Ireland: Decorative | |
| & Museum | B2 | | Arts & History | A2 |
| George's St Arcade | B2 | | | |

| Place | Grid |
|---|---|
| National Wax | |
| Museum Plus | B2 |
| Natural History | |
| Museum | C2 |
| Newman House | |
| Museum | B3 |
| Olympia Theatre | B2 |
| Park Shopping Centre | A1 |
| Pearse Station | C2 |
| Phoenix Care Centre | A1 |
| Project Arts Centre | B2 |
| Rotunda Hospital | B1 |
| Powerscourt Shopping | |
| Centre | B2 |
| Royal Victoria Eye & | |
| Ear Hospital | C3 |
| St. James's Hospital | A2 |
| St. Mary's | |
| Pro Cathedral | B2 |
| St. Patrick's Cathedral | B3 |
| St. Patrick's University | |
| Hospital | A2 |
| St. Stephen's Green | |
| Shopping Centre | B3 |
| Talbot Mall | B2 |
| Tara St Station | C2 |
| Technological University | |
| Dublin | A1 |
| The Ark | B2 |
| The CHQ Building | C2 |
| Trinity College | C2 |
| Vegetable Market | B2 |
| Vicar Street Theatre | A2 |
| Waterways Visitor | |
| Centre | C2 |

## General information

Population 1,173,179. The capital of the Republic of Ireland. The name is derived from the Irish 'Dubh Linn' meaning Black Pool. The Vikings originally settled on the south bank of the River Liffey and this is still the historic heart of the city.

## Tourist information

**Information lines**
☎ 01890 324 583 (Republic of Ireland)
☎ 0800 397000 (G.B & N. Ireland)

**Accommodation reservations**
☎ 1800 668 668 (Republic of Ireland)
☎ 0800 668 668 66 (G.B & N. Ireland)

The **Dublin Visitor Centres** are on O'Connell Street and Grafton Street ☎ 01 8980 700. They provide details of visitor attractions and events in the city. There is also a tourist information centre in Barnardo Square ☎ 01800 230 330. The official tourist information website for Dublin is www.visitdublin.com

The Four Courts                    Photo © HarperCollins Publishers

## Getting around

Dublin is linked with the cities and towns of Ireland by a network of rail and bus services overseen by **Córas Iompair Éireann (CIÉ)**, which is Ireland's National Transport Authority www.cie.ie

National bus services between Dublin, Dublin Airport and other major cities and towns are provided by **Irish Bus (Bus Éireann)** ☎ 01 836 6111 www.buseireann.ie

**Dublin Bus (Bus Átha Cliath)** operates the public bus services in Dublin and the surrounding area with its head office at 59 O'Connell Street Upper. ☎ 01 873 4222 www.dublinbus.ie

The two mainline railway stations, Dublin Connolly and Dublin Heuston, are operated by **Irish Rail (Iarnód Éireann)** ☎ 1850 366222 www.irishrail.ie. Irish rail also operates the suburban rail network in Dublin and DART (Dublin Area Rapid Transit) with stations between Malahide / Howth in the north to Greystones in the south.

**Luas** (the Irish for speed), the Dublin light railway system runs from Tallaght to Connolly station (with extensions to Saggart in the south west and Docklands / The Point in the east) and from Bride's Glen to Broombridge via St Stephen's Green ☎ 1800 300 604 www.luas.ie

Taxis are available at taxi ranks, including ones on O'Connell Street, Dame Street and St. Stephen's Green West.

## Places of interest

The places of interest section (pages 6-12) includes several places to visit within Dublin city centre - **Christ Church Cathedral, Dublin Castle**, the **National Museum of Ireland** and **St Patrick's Cathedral**. There are of course many more.

Landmark buildings include the fine **City Hall** building which was completed in 1779 as the Royal Exchange. Subsequent use included a prison and corn exchange before being taken over by the city in 1852. Another building of architectural interest is the **Custom House**, with its magnificent long river frontage. As well as being an iconic building the **General Post Office** on O'Connell Street was the headquarters of the 1016 uprising and its museum tells this story along with modern Irish history. Another museum is **EPIC The Irish Emigration Museum** which has video galleries, interactive touch screens, emigrant letters and an Irish Family History Centre. The **James Joyce Centre** is devoted to the great novelist and run by members of his family in a restored Georgian town house. The **Trinity College** complex of cobbled quadrangles and peaceful gardens has a library of over a million books, including the famous Book of Kells. Galleries include the **National Gallery of Ireland** with its Old Masters, illustrious 20th century European artists and the National Collection of Irish Art and the **Hugh Lane Municipal Art Gallery** which has an extensive collection of 19th and 20th century paintings, sculpture and stained glass.

## Theatres, concert halls & festivals

The **Abbey Theatre** ☎ 01 878 7222 is Ireland's National Theatre. Classic Irish plays are staged in the main theatre whilst the **Peacock Theatre** downstairs presents new and experimental drama. Other Dublin theatres include the **Gaiety Theatre** ☎ 01 646 8600 and the **Project Arts Centre** ☎ 01 881 9613. The **National Concert Hall**, Earlsfort Terrace ☎ 01 417 0000, is the home of the National Symphony Orchestra of Ireland and is also a venue for international artists and orchestras, jazz, contemporary and traditional Irish music.

Throughout the year Dublin hosts numerous festivals – the Film Festival in February, the St Patrick's Festival in March, the Writers Festival and Fringe Festival in September and the Theatre Festival held in late September to mid October.

## Shopping

The main city centre shopping areas are around Grafton Street and Nassau Street to the south of the Liffey and around Henry Street (off O'Connell Street) to the north of the river. Both Grafton Street and Henry Street are pedestrianised. Many up-market and international designer stores can be found in Grafton Street as well as the department stores of Arnotts and Brown Thomas. Shops in Henry Street include Dunnes department store. The rejuvenated cultural quarter of **Temple Bar** is the area to go for craft and specialist stores, as well as more alternative shops. Shopping centres include the **St Stephen's Green Centre**, the **Ilac Centre** in Henry Street, the **Jervis Shopping Centre** in Jervis Street and the **Powerscourt Centre** in South William Street.

## Parks & gardens

In the heart of the city **St Stephen's Green** was originally an open common but was enclosed in 1663. Opened to the general public in 1877, it is laid out as a public park with flowerbeds, an ornamental pond and several sculptures.

Northwest of the city centre **Phoenix Park** covers over 712 ha (1760 acres) and is Europe's largest enclosed city park. It was laid out in the mid 18th century and was the scene of the Phoenix Park murders in 1882, when the Chief Secretary and Under-Secretary for Ireland were assassinated. The park contains several buildings, including Áras an Uachtaráin, the official residence of the President of Ireland. In the southeast corner, by the main entrance, is **Dublin Zoo** and also the Victorian style ornamental planting of **The People's Garden**.

## Telephoning

If telephoning from the Republic of Ireland use the area code and telephone number. If telephoning from Great Britain or within Northern Ireland use the code 00353, delete the 0 from the area code and follow with the required telephone number.

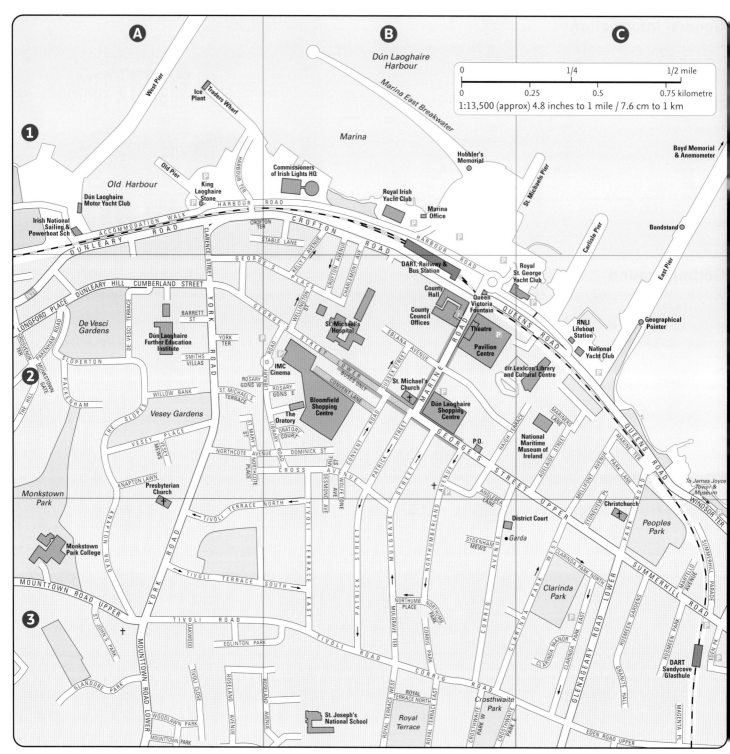

Dún Laoghaire (population 26,525) derives from the Irish 'Fort of Laoghaire', after an ancient king of Ireland. For a century the city was known as Kingstown until 1921 when it reverted to its Irish name. The town is sited on the coast about 7 miles (11km) south of Dublin.

## INDEX TO STREET NAMES

| | | | | | | | |
|---|---|---|---|---|---|---|---|
| Accommodation | | De Vesci Ter | A2 | Library Rd | B2 | Park La | C2 | Tivoli Rd | B3 |
| Walk | A1 | Desmond Ave | B2 | Longford Pl | A2 | Park Rd | C3 | Tivoli Ter East | B3 |
| Adelaide St | C2 | Dominick St | B2 | Magenta Pl | C3 | Patrick St | B2 | Tivoli Ter North | A3 |
| Anglesea La | B2 | Dunleary Hill | A2 | Marine Rd | B2 | Queens Rd | C2 | Tivoli Ter South | A3 |
| Barrett St | A2 | Dunleary Rd | A1 | Marine Ter | C2 | Rosary Gdns E | B2 | Vesey Pl | A2 |
| Charlemont Ave | B2 | Eblana Av | B2 | Mariners La | C2 | Rosary Gdns W | A2 | Vesey Mews | A2 |
| Clarence St | A1 | Eden Pk | C3 | Martello Av | A2 | Roseland Av | A3 | Wellington St | B2 |
| Clarinda Manor | C3 | Eden Rd Upper | C3 | Mellifont Av | C2 | Rosmeen Gdns | C3 | Willow Bank | A2 |
| Clarinda Park East | C3 | Eglinton Park | A3 | Mill St | B2 | Rosmeen Park | C3 | Windsor Ter | C2 |
| Clarinda Park North | C3 | George's Pl | B2 | Monkstown Gate | A2 | Royal Ter East | B3 | Woodlawn Park | A3 |
| Clarinda Park West | B2 | George's St Lwr | B2 | Mounttown Park | A3 | Royal Ter North | B3 | York Rd | A2 |
| Convent La | B2 | George's St Upr | B2 | Mounttown Rd Lwr | A3 | Royal Ter West | B3 | York Ter | A2 |
| Convent Rd | B2 | Glandore Park | A3 | Mounttown Rd Upr | A3 | St. John's Park | A3 | | |
| Corrig Av | B3 | Glenageary Rd Lwr | C3 | Mulgrave St | B3 | St. Mary's St | B2 | | |
| Corrig Park | B3 | Granite Hall | C3 | Mulgrave Ter | B3 | St. Michael's Ter | A2 | | |
| Corrig Rd | B3 | Grosvenor Ter | A2 | Northcote Av | A2 | Sloperton | A2 | | |
| Crofton Av | B2 | Haigh Ter | B2 | Northcote Pl | A2 | Slopes, The | A2 | | |
| Crofton Rd | B1 | Harbour Rd | B1 | Northumberland Av | B3 | Smiths Villas | A2 | | |
| Crofton Ter | A1 | Harbour Ter | A1 | Northumberland | | Stable La | B1 | | |
| Cross Av | B2 | Hill, The | A2 | Park | B3 | Stoneview Pl | C3 | | |
| Crosthwaite Park E | C3 | Kelly's Av | B2 | Northumberland Pl | B3 | Summerhill Parade | C3 | | |
| Crosthwaite Park W | B3 | Knapton Lawn | A3 | Oakwood | A3 | Summerhill Rd | C3 | | |
| Cumberland St | A2 | Knapton Rd | A3 | Oratory Court | B2 | Sussex St | B2 | | |
| | | | | Packenham | A2 | Sydenham Mews | B3 | | |
| | | | | Pakenham Rd | A2 | Tivoli Close | A3 | | |

## INDEX TO PLACES OF INTEREST

| | | | |
|---|---|---|---|
| Band Stand | C1 | Ice Plant | A1 |
| Bloomfield Shopping | | IMC Cinema | B2 |
| Centre | B2 | Irish National Sailing & | |
| Boyd Memorial and | | Powerboat School | A1 |
| Anemometer | C1 | King Laoghaire Stone | A1 |
| Carlisle Pier | C1 | Marina Office | B1 |
| Christchurch | C3 | Monkstown Park | A2 |
| Clarinda Park | C3 | Monkstown Park | |
| Club House | B3 | College | A3 |
| Commissioners of | | National Maritime | |
| Irish Lights HQ | B1 | Museum of Ireland | C2 |
| County Council | | National Yacht Club | C2 |
| Offices | B2 | Old Pier | A1 |
| County Hall | B2 | Pavilion Centre | B2 |
| Crosthwaite Park | B3 | Peoples Park | C3 |
| De Vesci Gardens | A2 | Presbyterian Church | A2 |
| DART Dún Laoghaire | B2 | Queen Victoria | |
| DART Sandycove | | Fountain | B2 |
| Glasthule | C3 | RNLI Lifeboat Station | C2 |
| dlr Lexlcon Library and | | Royal Irish Yacht Club | B1 |
| Cultural Centre | C2 | Royal St. George | |
| Dún Laoghaire Further | | Yacht Club | C2 |
| Education Institute | A2 | St. Michael's Church | B2 |
| Dún Laoghaire Motor | | St. Michael's Hospital | B2 |
| Yacht Club | A1 | St. Michaels Pier | C1 |
| Dún Laoghaire | | Swimming Pool | C2 |
| Shopping Centre | B2 | The Oratory | B2 |
| District Court | C3 | Traders Wharf | A1 |
| East Pier | C2 | Vesey Gardens | A2 |
| Geographical Pointer | C2 | West Pier | A1 |
| Hobbler's Memorial | B1 | | |

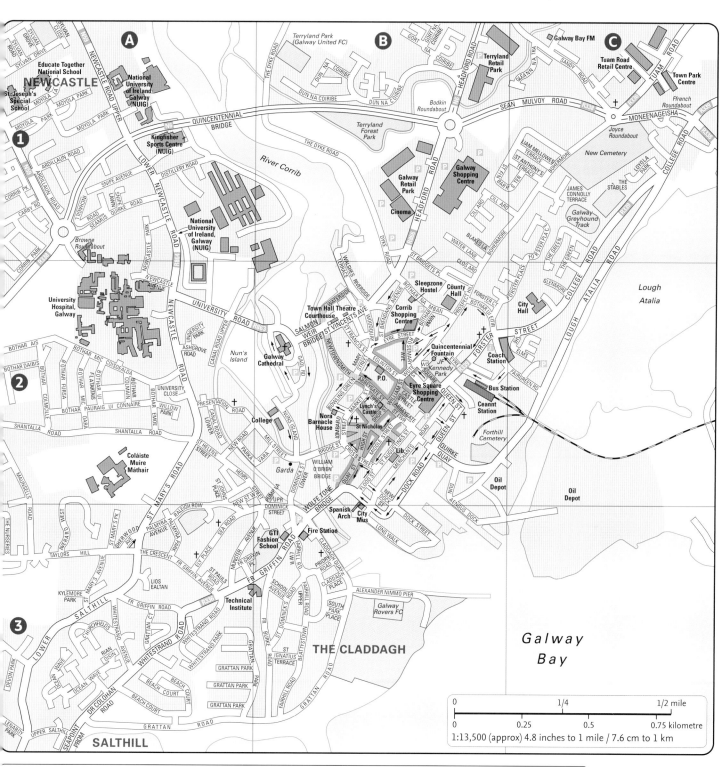

Galway (population 79,934) is one of the fastest growing cities in Europe and has seen enormous change over the last few years. Despite this it retains its intrinsic Irish character both physically, with the narrow winding streets of the city centre, and culturally by being on the edge of one of the largest Gaeltacht areas in Ireland.

Scale: 0 — 1/4 — 1/2 mile; 0 — 0.25 — 0.5 — 0.75 kilometre
1:13,500 (approx) 4.8 inches to 1 mile / 7.6 cm to 1 km

## INDEX TO STREET NAMES

| | | | | | |
|---|---|---|---|---|---|
| Abbeygate St Lwr | B2 | Bridge St | B2 | Eyre Sq | B2 |
| Abbeygate St Upr | B2 | Canal Rd Lwr | A2 | Eyre St | B2 |
| Alexander Nimmo Dr | B3 | Canal Rd Upr | A2 | Fairgreen Rd | B2 |
| Ardillaun Rd | A1 | Carby Rd | A1 | Fairhill Rd | B3 |
| Averard West | A3 | Castle St | A2 | Fairhill Rd Lwr | B3 |
| Beach Ct | A3 | Cill Ard | B1 | Fairhill Rd Upr | B3 |
| Beattystown | B3 | Claddagh Pl | B3 | Father Burke Rd | B1 |
| Blakes La | B1 | Claddagh Quay | B3 | Father Griffin Av | A3 |
| Bohermore | B1 | Clos Ard | B1 | Father Griffin Pl | A3 |
| Bóthar Ais | A2 | College Rd | C1 | Father Griffin Rd | A3 |
| Bóthar Colmain | A2 | College Rd | C2 | Flood St | B2 |
| Bóthar Colmcille | A2 | Corrib Park | A1 | Forster Ct | C2 |
| Bóthar Daibis | A2 | Crescent, The | A3 | Forster Pl | B2 |
| Bóthar Emoe | A2 | Cross St | B2 | Forster St | B2 |
| Bóthar Fursa | A2 | Devon Pk | A3 | Gaol Rd | B2 |
| Bóthar Irwin | A2 | Distillery Rd | A1 | Gleann na Tra | C1 |
| Bóthar Mic Coisdealda | A2 | Dock Rd | B2 | Glenmore | C2 |
| Bóthar Mic Oara | A2 | Dock St | B3 | Gort na Coiribe | B1 |
| Bóthar na Mban | B2 | Doctor Colohan Rd | A3 | Grattan Ct | A3 |
| Bóthar Pauraig Ui Connaire | A2 | Dominick St Lwr | A2 | Grattan Pk | A3 |
| Bóthar Ui Eitir | B2 | Dun Aengus Dock | B2 | Grattan Rd | A3 |
| Bóthar Ui Flatartais | B1 | Dun na Coiribe | B1 | Green, The | C1 |
| Bowling Green | B2 | Dyke Rd, The | B1 | Headford Rd | B1 |
| | | Eglinton St | B2 | Henry St | A2 |
| | | | | High St | B2 |

| | | | | | |
|---|---|---|---|---|---|
| Elms, The | C2 | Inchagill Rd | A1 | Newcastle Rd | A2 |
| Ely Pl | A3 | James Connolly Ter | C1 | Newcastle Rd Upr | A1 |
| Eyre Sq | B2 | Kylemore Pk | A3 | Newtownsmith | B2 |
| Eyre St | B2 | Lenaboy Pk | A3 | Nuns Island | B2 |
| Fairgreen Rd | B2 | Liam Mellowes Ter | C1 | Nuseries, The | A3 |
| Fairhill Rd | B3 | Lios Ealtan | A3 | Ocean Wave | A3 |
| Fairhill Rd Lwr | B3 | Lombard St | B2 | Palmyra Av | B2 |
| Fairhill Rd Upr | B3 | Long Walk | B3 | Palmyra Pk | A3 |
| Father Burke Rd | B1 | Lough Atalia Rd | C1 | Parka Vara | A2 |
| Father Griffin Av | A3 | Lower Newcastle Rd | A1 | Presentation Rd | A2 |
| Father Griffin Pl | A3 | Lower Salthill | A3 | Priory Rd | B3 |
| Father Griffin Rd | A3 | Loyola Pk | A3 | Prospect Hill | C1 |
| Flood St | B2 | Market St | B2 | Pump La | B2 |
| Forster Ct | C2 | Mary St | B2 | Quay St | B2 |
| Forster Pl | B2 | Maunsells Rd | A2 | Queen St | B2 |
| Forster St | B2 | Merchants Rd | B2 | Quincentennial Bridge | A1 |
| Gaol Rd | B2 | Middle St | B2 | Quirke Quay | B2 |
| Gleann na Tra | C1 | Mill St | B2 | Raleigh Row | A3 |
| Glenmore | C2 | Moneenageisha | C1 | Rian Oisin | A3 |
| Gort na Coiribe | B1 | Moyola La | A1 | Riverside | A1 |
| Grattan Ct | A3 | Moyola Pk | A1 | Rosemary Av | B2 |
| Grattan Pk | A3 | Munster Av | A3 | St. Anthony's Ter | C1 |
| Grattan Rd | A3 | New Dock St | B2 | St. Augustines St | B2 |
| Green, The | C1 | New Rd | A2 | St. Brendans Av | B2 |
| Headford Rd | B1 | New St West | A2 | St. Bridgets Pl | B1 |
| Henry St | A2 | Newcastle Av | A2 | St. Dominick's Rd | B3 |
| High St | B2 | Newcastle Pk | A2 | St. Finbarr's Ter | B1 |

| | | | | | |
|---|---|---|---|---|---|
| St. Francis St | B2 | Taylors Hill | A3 |
| St. Helen's St | A1 | Tuam Rd | C1 |
| St. Ignatius Ter | B3 | University Cl | A2 |
| St. John's Pl | A2 | University Pk | A2 |
| St. Mary's Av | A3 | University Rd | A2 |
| St. Mary's Pk | A3 | Upper Dominick St | B2 |
| St. Mary's Rd | A3 | Upper Salthill | A3 |
| St. Pauls Rd | A3 | Walshes Ter | B1 |
| St. Vincents Av | B2 | Water La | B1 |
| Salmon Weir Bridge | B2 | Waterside | B2 |
| Sandy Rd | C1 | Wellpark | C1 |
| School Av | B2 | Whitestrand Av | A3 |
| Sea Rd | A3 | Whitestrand Pk | A3 |
| Seamus Quirke Rd | A1 | Whitestrand Rd | A3 |
| Sean Mulvoy Rd | B1 | William St | B2 |
| Seapoint Promenade | A3 | Williamsgate St | B2 |
| Shantalla Rd | A2 | Willow Pk | A2 |
| Sherwood Av | A3 | Wolfe Tone Bridge | B3 |
| Shop St | B2 | Woodquay | B2 |
| Snipe Av | A1 | | |
| Snipe Lawn | A1 | | |
| South Park Pl | B3 | | |
| Stables, The | C1 | | |
| Sylvan Cl | A1 | | |
| Sylvan Dr | A1 | | |
| Sylvan Gro | A1 | | |
| Sylvan Rd | A1 | | |

## INDEX TO PLACES OF INTEREST

| | |
|---|---|
| Ceannt Railway Station | B2 |
| Cinema | B1 |
| City Hall | C2 |
| City Museum | B3 |
| Coach Station | B2 |
| Corrib Shopping Centre | B2 |
| County Hall | B2 |
| Eyre Square Shopping Centre | B2 |
| Galway Bay FM | C1 |
| Galway Cathedral | B2 |
| Galway City Museum | B2 |
| Galway Retail Park | B1 |
| Galway Shopping Centre | B1 |
| J F Kennedy Park | B2 |
| Kingfisher Sports Centre (NUIG) | A1 |
| Lynch's Castle | B2 |
| National University of Ireland, Galway (NUIG) | A1 |
| Nora Barnacle House | B2 |
| Quincentennial Fountain | B2 |
| Sleepzone Hostel | B2 |
| Spanish Arch | B3 |
| Terryland Forest Park | B1 |
| Terryland Retail Park | B1 |
| Town Hall Theatre | B2 |
| Town Park Centre | C1 |
| Tuam Road Retail Centre | C1 |
| University Hospital, Galway | A2 |

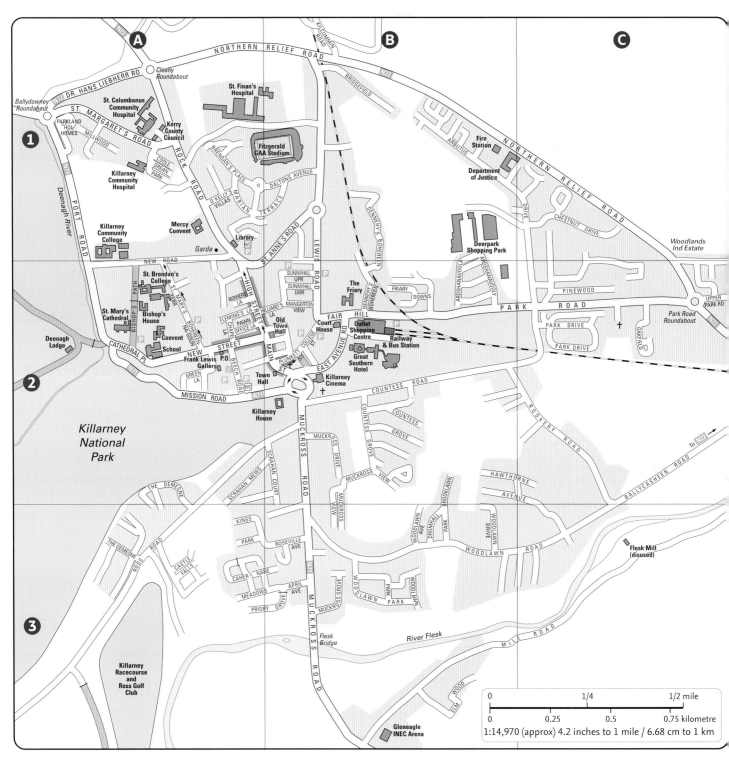

Killarney (population 14,504) is set on the edge of the beautiful Killarney National Park, near the shores of Lough Leane, and is one of Ireland's most popular tourist centres. An ideal base to explore the beautiful peninsulas of Dingle and Kerry, home to Ireland's highest mountain, Carrantuohill in the Macgillycuddy's Reeks.

## INDEX TO STREET NAMES

| | | | | | | | |
|---|---|---|---|---|---|---|---|
| April Av | **B3** | Coolegrean Pk | **A1** | Kings Pk | **A3** | Pinewood | **C2** |
| Arbutus Dr | **B1** | Countess Gr | **B2** | Lewis Rd | **B1** | Plunkett St | **B2** |
| Ardshanavooly | **B2** | Countess Rd | **B2** | Main St | **B2** | Port Rd | **A1** |
| Ballycasheen Rd | **C2** | Daltons Av | **B1** | Mangerton View | **B2** | Priory Gr | **A3** |
| Beech Rd | **A2** | Dennehy's | | Marian Ter | **A1** | Rock Rd | **A1** |
| Bishop's Path | **A2** | Bohereen | **B1** | Mill Rd | **B3** | Rookery Rd | **C2** |
| Bohereen-Na- | | Dr. Hans | | Millwood | **A1** | Roseville Av | **B3** |
| Goun | **A2** | Liebherr Rd | **A1** | Mission Rd | **A2** | Ross Rd | **A3** |
| Bowers La | **A2** | Dromhall Pk | **B3** | Muckross Dr | **B2** | St. Anne's Rd | **B1** |
| Brendan's Pl | **A1** | East Av Rd | **B2** | Muckross Gr | **B3** | St. Margaret's Rd | **A1** |
| Bridgefield | **B1** | Elm Wood | **B3** | Muckross Rd | **B2** | St. Mary's Terr | **A2** |
| Brooklawn | **B2** | Fair Hill | **B2** | Muckross View | **B2** | Scrahan Ct | **B2** |
| Cahernane | | Fleming's La | **A2** | New Rd | **A1** | Scrahan Mews | **A2** |
| Meadows | **A3** | Friary Downs | **B2** | New Street | **A2** | Sunnyhill Lwr | **B2** |
| Castle Falls | **A3** | Green La | **A2** | Northern | | Sunnyhill Upr | **B2** |
| Cathedral Pl | **A2** | Hawthorne Av | **B2** | Relief Rd | **A1** | The Demesne | **A2** |
| Chapel La | **A2** | High Street | **A2** | Oakfield | **C2** | Upper Pk Rd | **C2** |
| Chestnut Dr | **C1** | Hilliard's La | **B2** | O'Kelly's Villas | **A1** | Woodlawn Av | **B3** |
| College St | **B2** | Kilcummin Rd | **B1** | Park Dr | **C2** | Woodlawn Dr | **B3** |
| | | | | Park Rd | **B2** | Woodlawn Pk | **B3** |
| | | | | Pawn Office La | **A2** | Woodlawn Rd | **B3** |

## INDEX TO PLACES OF INTEREST

| | | | |
|---|---|---|---|
| Bishop's House | **A2** | Killarney | |
| Court House | **B2** | Racecourse | **A3** |
| Deenagh Lodge | **A2** | Malton Hotel | **B2** |
| Deerpark Shopping | | Mercy Convent | **A1** |
| Park | **B1** | Old Town Hall | **B2** |
| Department of | | Outlet Shopping | |
| Justice | **B1** | Centre | **B2** |
| Fire Station | **B1** | Railway & Bus | |
| Fitzgerald GAA | | Station | **B2** |
| Stadium | **B1** | Ross Golf Club | **A3** |
| Flesk Mill | | St. Brendan's | |
| (disused) | **C3** | College | **A2** |
| Frank Lewis | | St. Columbanus | |
| Gallery | **A2** | Community | |
| Gleneagle INEC | | Hospital | **A1** |
| Arena | **B3** | St. Finan's | |
| Kerry County | | Hospital | **A1** |
| Council | **A1** | St. Mary's | |
| Killarney Cinema | **B2** | Cathedral | **A2** |
| Killarney Community | | The Friary | **B2** |
| College | **A1** | Town Hall | **B2** |
| Killarney Community | | | |
| Hospital | **A1** | | |
| Killarney House | **B2** | | |
| Killarney National | | | |
| Park | **A2** | | |

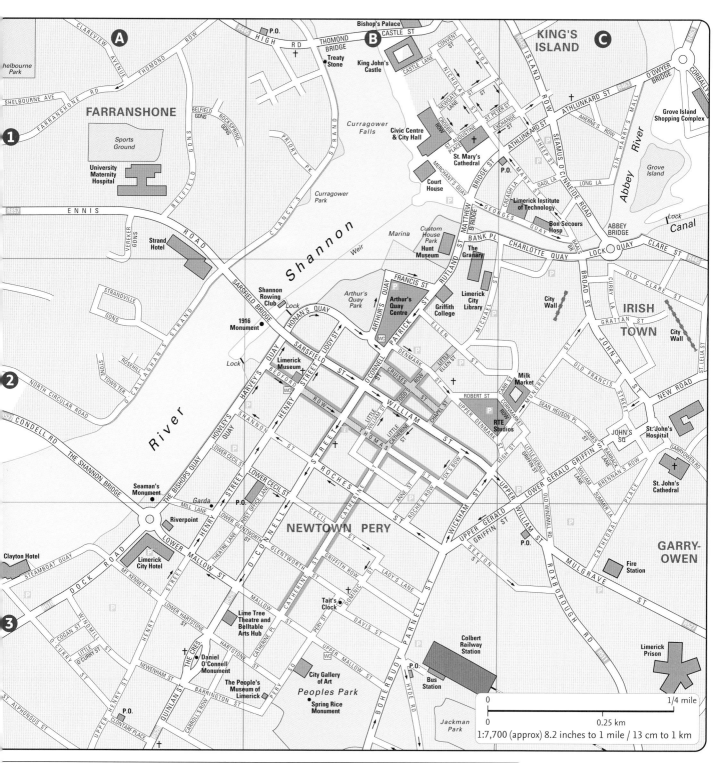

Limerick (population 94,192) lies at a strategic point on the River Shannon. The city is a major regional centre and is often considered to be Ireland's sporting, especially rugby, capital. The high kick followed by the charge to recapture the ball is named after the Garryowen Rugby Club, to the south west of Limerick, alleged to have been the first to use it.

## INDEX TO STREET NAMES

| | | | |
|---|---|---|---|
| ...ey Bridge | C1 | Clontarf Pl | A3 |
| ...rne's Row | C1 | Cogan St | A3 |
| ...e St | B2 | Condell Rd | A2 |
| ...hur's Quay | B2 | Convent St | B1 |
| ...lunkard St | C1 | Cornmarket Row | C2 |
| ...ls Bridge | C1 | Creagh La | B1 |
| ...ck Pl | B1 | Crosbie Row | B1 |
| ...rack La | C2 | Cruises St | B2 |
| ...ford Row | B2 | Curry La | C2 |
| ...rington St | A3 | Davis St | B3 |
| ...field Gdns | A1 | Denmark St | B2 |
| ...hop St | B1 | Dock Rd | A3 |
| ...erbuoy | B3 | Dominic St | B3 |
| ...nnan's Row | C2 | Ellen St | B2 |
| ...ge St | B1 | Ennis Rd | A1 |
| ...ad St | C2 | Exchange St | C2 |
| ...e St | C2 | Farranshone Rd | A1 |
| ...oll's Row | A3 | Fox's Bow | B2 |
| ...cle St | B1 | Francis St | B1 |
| ...hedral Pl | C3 | Gaol La | C1 |
| ...therine Pl | B3 | Garryowen Rd | C2 |
| ...herine St | B3 | Georges Quay | B1 |
| ...l St | B3 | Glentworth St | B3 |
| ...pel St | B2 | Grattan St | C2 |
| ...rlotte Quay | C1 | Griffith Row | B3 |
| ...ncy's Strand | B1 | Hartstone St | A3 |
| ...se St | C1 | Harvey's Quay | A2 |
| ...eview Av | A1 | Henry St | A3 |

| | | | |
|---|---|---|---|
| High Rd | B1 | | |
| High St | C2 | | |
| Honan's Quay | B2 | | |
| Howley's Quay | A2 | | |
| Hyde Rd | B3 | | |
| Island Row | C1 | | |
| James St | C2 | | |
| John's Sq | C2 | | |
| John's St | C2 | | |
| Lady's La | B3 | | |
| Liddy St | B2 | | |
| Little Catherine St | B2 | | |
| Little Ellen St | B2 | | |
| Little Gerald Griffin St | B2 | | |
| Little O'Curry St | A3 | | |
| Little William St | B2 | | |
| Lock Quay | C1 | | |
| Long La | C1 | | |
| Lower Cecil St | A2 | | |
| Lower Gerald Griffin St | B2 | | |
| Lower Glentworth St | A3 | | |
| Lower Hartstone St | A3 | | |
| Lower Mallow St | B3 | | |
| Mallow St | B3 | | |
| Mary St | C2 | | |
| Matthew Bridge | B1 | | |
| Merchant's Quay | B1 | | |
| Michael St | B2 | | |
| Mill La | A3 | | |
| Mt Kennett Pl | A3 | | |

| | | | |
|---|---|---|---|
| Mulgrave St | C3 | St. Frances St | B1 |
| Mungret St | C2 | St. Lelia St | C2 |
| New Rd | C2 | St. Peter St | B1 |
| Newenham St | A3 | Sarsfield Bridge | A2 |
| Newgate La | B1 | Sarsfield St | B2 |
| Nicholas St | B1 | Seamus O'Cinneide Rd | C1 |
| North Circular Rd | A2 | Sean Heuson Pl | C2 |
| O'Callaghan's Strand | A2 | Sexton St | B3 |
| O'Connell St | B2 | Shannon St | A2 |
| O'Curry St | B3 | Sheep St | C1 |
| O'Dwyer Bridge | C1 | Shelbourne Ave | A1 |
| Old Clare St | C1 | Sir Harry's Hall | C1 |
| Old Francis St | C2 | Steamboat Quay | A3 |
| Old Windmill Rd | B3 | Stonetown Ter | A2 |
| Parnell St | B3 | Strandville Gdns | A2 |
| Patrick St | B2 | Summer La | C2 |
| Pery Sq | B3 | The Bishops Quay | A2 |
| Pery St | B3 | The Crescent | A3 |
| Post Office La | B3 | The Shannon Bridge | A2 |
| Priory Pk | B1 | Theatre La | A3 |
| Quinlan St | A3 | Thomas St | B2 |
| Robert St | B2 | Thomond Bridge | B1 |
| Roches Row | B3 | Thomond Row | B1 |
| Roches St | B2 | Todd's Row | B2 |
| Rockspring Gdns | C1 | Upper Denmark St | B2 |
| Rosehill | A2 | Upper Gerald Griffin St | B3 |
| Roxborough Rd | C3 | Upper Henry St | A3 |
| Rutland St | B2 | Upper Mallow St | B3 |
| St. Alphonsus St | A3 | Upper William St | C2 |
| St. Augustine Pl | B1 | Vereker Gdns | A1 |

| | | | |
|---|---|---|---|
| Wickham St | B3 | | |
| William La | C2 | | |
| William St | B2 | | |
| Windmill St | A3 | | |

## INDEX TO PLACES OF INTEREST

| | | | |
|---|---|---|---|
| 1916 Monument | B2 | St. John's Cathedral | C2 |
| Arthur's Quay Centre | B2 | St. John's Hospital | C2 |
| Arthur's Quay Park | B2 | St. Mary's Cathedral | B1 |
| Bishop's Palace | B1 | St. Mary's Town House | C1 |
| Bon Secours Hosptial | C1 | Seaman's Monument | A2 |
| Bus Station | B3 | Shannon Rowing Club | A2 |
| City Gallery of Art | B3 | Spring Rice Monument | B3 |
| City Wall | C2 | Strand Hotel | A1 |
| Civic Centre & City Hall | B1 | Tait's Clock | B3 |
| Clayton Hotel | A3 | The Granary | B2 |
| Colbert Railway | | The People's Museum of | |
| Station | B3 | Limerick | B3 |
| Court House | B1 | Treaty Stone | B1 |
| Curragower Park | B1 | University Maternity | |
| Custom House Park | B1 | Hospital | A1 |
| Daniel O'Connell | | | |
| Monument | A3 | | |
| Griffith College | B2 | | |
| Grove Island Shopping | | | |
| Complex | C1 | | |
| Hunt Museum | B1 | | |
| Jackman Park | B3 | | |
| King John's Castle | B1 | | |
| Limerick City Hotel | A3 | | |
| Limerick City Library | B2 | | |
| Limerick Institute of | | | |
| Technology | C1 | | |
| Limerick Museum | B1 | | |
| Limerick Prison | C3 | | |
| Lime Tree Theatre and | | | |
| Belltable Arts Hub | A3 | | |
| Milk Market | C2 | | |
| Peoples Park | B3 | | |
| Post Office | B1 | | |
| RTE Studios | C3 | | |

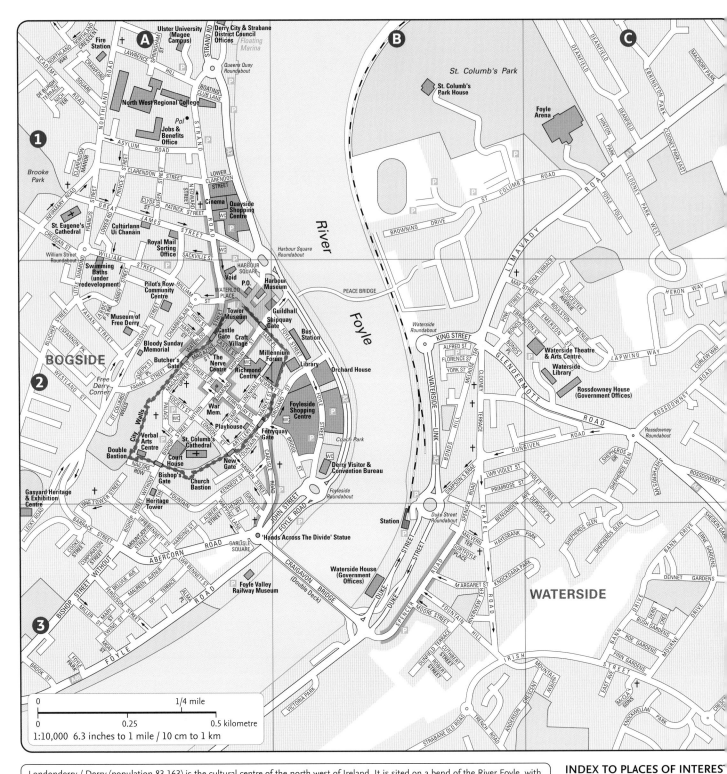

Londonderry / Derry (population 83,163) is the cultural centre of the north west of Ireland. It is sited on a bend of the River Foyle, with historic Cityside on the west bank and Waterside on the east bank and is the only completely walled city in Ireland. Four main streets radiate from the Diamond to four gateways set in the 17th century walls which have withstood many sieges in the city's turbulent past.

## INDEX TO STREET NAMES

| | | | | | |
|---|---|---|---|---|---|
| Abbey Pk | A2 | Carlisle Sq | A3 | Ebrington Pk | C1 |
| Abbey St | A2 | Castle St | A2 | Ebrington St | B2 |
| Abercorn Rd | A3 | Chamberlain St | A2 | Emerson St | C2 |
| Academy Rd | A1 | Chapel Rd | B3 | Erne Gdns | C3 |
| Alfred St | B2 | Clarendon Manor | A1 | Ewing St | A3 |
| Alma Pl | A3 | Clarendon St | A1 | Fahan St | A2 |
| Anderson Cres | B3 | Clooney Pk East | C1 | Ferguson St | A2 |
| Artillery St | A2 | Clooney Pk West | C1 | Ferryquay St | A2 |
| Asylum Rd | A1 | Clooney Ter | B2 | Finn Gdns | C3 |
| Aubery St | A3 | Clyde St | A1 | Florence St | B2 |
| Bank Pl | A2 | Cooke St | A3 | Fountain Hill | B3 |
| Bann Dr | C3 | Corporation St | A1 | Fountain St | A2 |
| Barrack St | A3 | Craigavon Bridge | B3 | Fountain, The | A2 |
| Bellvue Av | A3 | Crawford Sq | A1 | Foyle Fold | C1 |
| Benvarden Av | B3 | Creggan St | A1 | Foyle Pk | A3 |
| Bishop St Within | A2 | Crescent Link | C3 | Foyle Rd | A3 |
| Bishop St Without | A3 | Curlew Way | C2 | Foyle St | A2 |
| Blucher St | A2 | Cuthbert St | B3 | Francis St | A1 |
| Boating Club La | A1 | De Burgh Ter | A1 | Glendermott Rd | B2 |
| Bonds Pl | B2 | Deanfield | C1 | Gloucester Av | C1 |
| Bonds St | B2 | Dennet Gdns | C3 | Gortfoyle Pl | B3 |
| Bonds St | B2 | Derg Cres | C3 | Grove Pl | A2 |
| Bridge St | B2 | Dervock Pl | A3 | Great James St | A1 |
| Brook St | A3 | Diamond, The | A2 | Harbour Sq | B2 |
| Browning Dr | B1 | Duke St | B3 | Harding St | A3 |
| Bush Gdns | C3 | Dunfield Ter | B3 | Harvey St | A2 |
| Butcher St | A2 | Dungiven Rd | C2 | Hawkin St | A3 |
| Carlisle Rd | A2 | East Av | C3 | Hayesbank Pk | B3 |

| | | | | | |
|---|---|---|---|---|---|
| High St | A2 | Market St | A2 | Rossdowney Pk | C2 |
| Hinton Pk | B2 | Mary St | A3 | Rossdowney Rd | C2 |
| Infirmary Rd | A1 | Maureen Av | A3 | Rossville St | A2 |
| Iona Ter | C2 | May St | B2 | Roulston Av | C2 |
| Irish St | B3 | McClay Gdns | C3 | Sackville St | A1 |
| Ivy Ter | A3 | Melrose Ter | B2 | St. Columb's Rd | B1 |
| John St | B3 | Miller St | A3 | St. Columbs Wells | A2 |
| Joseph's Pl | A2 | Moat St | A2 | Shepherds Glen | C2 |
| Kennedy St | A2 | Moore St | B3 | Shepherds Grove | C2 |
| King St | B2 | Mountainview | B2 | Shepherds Way | C2 |
| Knockdara Pk | B3 | Mountjoy St | A3 | Sheskin Gdns | C3 |
| Knockwellan Pk | C3 | Mourne Dr | C3 | Shipquay St | A2 |
| Lapwing Way | C2 | Nailors Row | A2 | Simpson's Brae | B2 |
| Lawrence Hill | A1 | Newmarket St | A2 | Society St | A2 |
| Lecky Rd | A3 | Nicholson Ter | A1 | Spencer Rd | B3 |
| Limavady Rd | B2 | North Edward St | A1 | Springham St | A1 |
| Linenhall St | A2 | Northland Cres | A1 | Strabane Old Rd | B3 |
| Lisfannon Pk | A2 | Northland Rd | A1 | Strand Rd | A1 |
| London St | A2 | Northland Way | A1 | Trench Rd | B3 |
| Long Tower St | A3 | Orchard St | B2 | Upper Bennett St | A3 |
| Lower Bennett St | A3 | Palace St | A2 | Violet St | A2 |
| Lower Clarendon St | A1 | Patrick St | A1 | Wapping La | A2 |
| Lower Rd | A1 | Pine St | B2 | Waterloo St | A2 |
| Lower Violet St | B2 | Primrose St | A1 | Waterside Link | B2 |
| Macrory Pk | C1 | Prince's St | A1 | Westland St | A2 |
| Magazine St | A2 | Pump St | A2 | William St | A1 |
| Magazine St Upper | A2 | Queen St | A2 | York St | B2 |
| Malvern Ter | B3 | Riverview Ter | B3 | | |
| Margaret St | B3 | Robert St | B3 | | |
| | | Roe Gdns | C3 | | |

## INDEX TO PLACES OF INTERES[T]

| | | | |
|---|---|---|---|
| Bishop's Gate | A2 | Orchard House | B2 |
| Bus Station | B2 | Peace Bridge | B2 |
| Butcher's Gate | A2 | Pilot's Row Community | |
| Castle Gate | A2 | Centre | A2 |
| Church Bastion | A2 | Playhouse | A2 |
| Cinema | A1 | Post Office | A2 |
| City Walls | A2 | Quayside Shopping | |
| Court House | A2 | Centre | A1 |
| Craft Village | A2 | Richmond Centre | A2 |
| Cultúrlann Uí Chanáin | A1 | Rossdowney House | |
| Derry City & Strabane | | (Government | |
| District Council Off. | A1 | Offices) | C2 |
| Derry Visitor & | | Royal Mail Sorting | |
| Convention Bureau | B2 | Office | A1 |
| Double Bastion | A2 | St. Columb's Cathedral | A2 |
| Ferryquay Gate | A2 | St. Columb's Park | C1 |
| Fire Station | A1 | St. Columb's | |
| Foyle Arena | C1 | Park House | B1 |
| Foyle Valley Railway | | St. Eugene's Cathedral | A1 |
| and Museum | A3 | Shipquay Gate | B2 |
| Foyleside Shopping Cen | B2 | Swimming Baths | A2 |
| Gasyard Heritage & | | The Nerve Centre | A2 |
| Exhibition Centre | A2 | Tower Museum | A2 |
| Guildhall | B2 | Ulster University | |
| 'Hands Across The Divide' | | (Magee Campus) | A1 |
| Statue | A3 | Verbal Arts Centre | A2 |
| Harbour Museum | B2 | Void (Art Gallery) | A2 |
| Heritage Tower | A2 | War Memorial | A2 |
| Library | B2 | Waterside House | |
| Millennium Forum | B2 | (Government | |
| Museum of Free Derry | A2 | Offices) | B3 |
| New Gate | A2 | Waterside Library | B2 |
| North West Regional | | Waterside Theatre & | |
| College | A1 | Arts Centre | B2 |

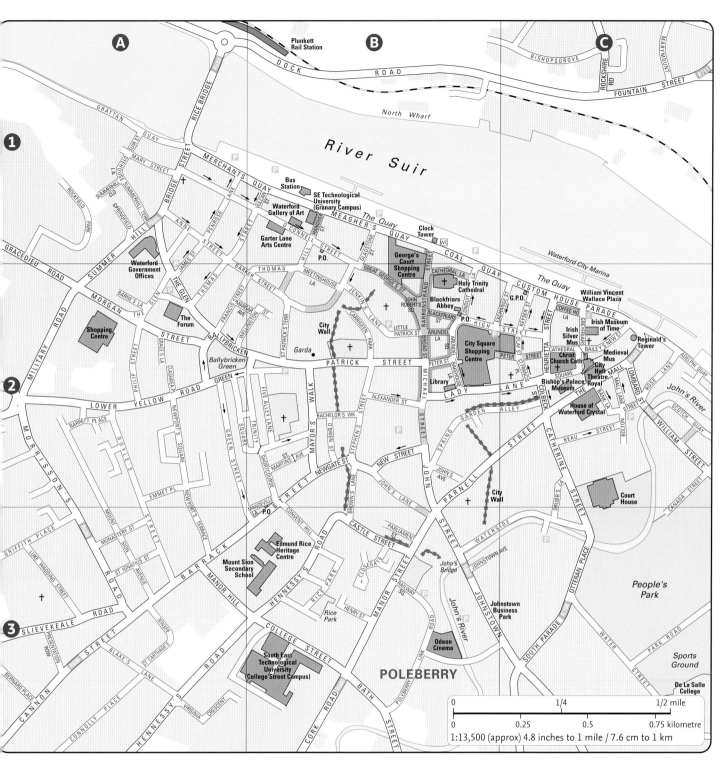

Waterford (population 53,504) dates from Viking times and developed as a port on the estuary of the River Suir. It is still the largest port in south east Ireland, as well as a major commercial centre. Waterford is best known for Waterford Crystal, which started in the city in 1783. The 'House of Waterford Crystal' opened in the centre of Waterford in 2010.

## INDEX TO STREET NAMES

| | | | | | |
|---|---|---|---|---|---|
| delphi Quay | C2 | Francis St | A2 | Lombard St | C2 |
| exander St | B2 | Gladstone St | B1 | Lower Newtown Rd | C3 |
| nne St | A1 | Gracedieu Rd | A1 | Lower Yellow Rd | A2 |
| rundel La | B2 | Grace's La | A2 | Luke Wadding St | A3 |
| rundel Sq | B2 | Grattan Quay | B2 | Mall La | C2 |
| achelor's Wk | B2 | Great George's St | B2 | Mandeville La | B2 |
| ailey's New St | C2 | Green St | A2 | Manor Hill | A3 |
| akehouse La | B2 | Greyfriars | C2 | Manor St | B3 |
| allybricken Green | A2 | Griffith Pl | A3 | Margar Av | A2 |
| ank La | C2 | Grubb's La | C2 | Mary St | A1 |
| arker St | A1 | Hanover St | B1 | Marymount | C1 |
| arne's La | A2 | Hennessy's Rd | A3 | Mayor's Walk | B2 |
| arrack St | A3 | Henrietta St | C2 | Meagher's Quay | B1 |
| arrett Pl | A2 | Henry St | A3 | Meetinghouse La | B2 |
| arronstrand St | B2 | High St | B2 | Merchants Quay | A1 |
| ath St | B3 | James St | A2 | Michael St | B2 |
| eau St | C2 | Jenkin's La | B2 | Military Rd | A2 |
| ernard Pl | A3 | John Robert's Sq | B2 | Monastery St | A3 |
| ishopsgrove | C1 | John St | B2 | Morgan St | A2 |
| ackfriars St | B2 | John's Hill | C3 | Morrisson's Rd | A2 |
| ake's La | A3 | John's La | B2 | Mount Sion Av | A3 |
| olton St | C2 | John's Ave | B2 | New St | B2 |
| ridge St | A1 | Johnstown | B3 | Newgate St | B2 |
| rown's La | B2 | Johnstown Av | B3 | Newport's Sq | A2 |
| | | Keyser St | C2 | Newport's Ter | A2 |
| | | Lady La | B2 | O'Brien St | B2 |
| | | Little Patrick St | B2 | O'Connell St | B1 |

| | | | | | |
|---|---|---|---|---|---|
| Canada St | C2 | Olaf St | C2 | Springfield | A1 |
| Cannon St | A3 | Otteran Pl | C3 | Stephen's St | B2 |
| Carrigeen Pk | B2 | Park Rd | C3 | Suir St | A1 |
| Castle St | B3 | Parliament St | B3 | Summer Hill | A2 |
| Cathedral La | B2 | Parnell St | B2 | Summerhill Cl | A1 |
| Catherine St | C2 | Patrick St | B2 | Summerhill Terr | A1 |
| Clashrea Pl | A2 | Penrose La | A1 | The Glen | A2 |
| Closegate | B3 | Peter St | B2 | The Mall | C2 |
| Coal Quay | B1 | Philip St | A2 | Thomas Hill | A2 |
| Coffee Ho La | C2 | Poleberry Link Road | B3 | Thomas St | A2 |
| College St | B3 | Presentation Row | A3 | Trinity Sq | B2 |
| Connolly Pl | A3 | Railway Sq | B3 | Virginia Cres | B3 |
| Convent Hill | B2 | Rice Bridge | A1 | Vulcan St | B1 |
| Cork Rd | C2 | Rice Pk | B3 | Water St | C3 |
| Custom House Parade | C2 | Rockfield Pk | A1 | Waterside | B3 |
| Dock Rd | B1 | Rockshire Rd | C1 | Wellington St | A2 |
| Doyle's St | A2 | Rose La | C2 | William St | C2 |
| Dye House La | A1 | St Carthage's Av | A3 | | |
| Emmet Pl | A2 | St Francis St | C2 | | |
| Exchange St | C2 | St Ignatius St | A3 | | |
| Five Alley La | B2 | St Martins Ave | A3 | | |
| Fountain St | C1 | St Patrick's Terr | B2 | | |
| | | Scotch Quay | B2 | | |
| | | Shortcourse | B2 | | |
| | | Slievekeale Rd | A3 | | |
| | | South Parade | B3 | | |
| | | Spring Garden Alley | B2 | | |

## INDEX TO PLACES OF INTEREST

| | | | |
|---|---|---|---|
| Ballybricken Green | A2 | Odeon Cinema | B3 |
| Bishop's Palace Museum | B2 | People's Park | C3 |
| Blackfriars Abbey | B2 | Plunkett Bus and Rail Station | B1 |
| Bus Station | B1 | Reginald's Tower | C2 |
| Christ Church Cathedral | C2 | Shopping Centre | A2 |
| City Hall | C2 | South East Technological University (College Street Campus) | B3 |
| City Square Shopping Centre | B2 | | |
| City Wall | B2 | South East Technological University (Granary Campus) | B1 |
| Clock Tower | B1 | | |
| Court House | C2 | The Forum | A2 |
| De La Salle College | C3 | Theatre Royal | C2 |
| Edmund Rice Heritage Centre | B2 | Waterford Gallery of Art | B1 |
| Garter Lane Arts Centre | B1 | Waterford Government Offices | A1 |
| General Post Office | C2 | William Vincent Wallace Plaza | C2 |
| George's Court Shopping Centre | B1 | | |
| Holy Trinity Cathedral | B2 | | |
| House of Waterford Crystal | C2 | | |
| Irish Museum of Time | C2 | | |
| Irish Silver Museum | C2 | | |
| Johnstown Business Park | C3 | | |
| Library | B2 | | |
| Medieval Museum | C2 | | |

## A

Aasleagh **25** D1
Abbey **31** D2
Abbeydorney **35** D3
Abbeyfeale
  (Mainistir na Feile) **35** E3
Abbeyknockmoy **26** B3
Abbeylara **28** A2
Abbeyleix
  (Mainistir Laoise) **32** B3
Abbeyshrule **28** A3
Abington **37** D2
Acaill (Achill) **18** C4
Achill (Acaill) **18** C4
Achill Sound
  (Gob an Choire) **18** C4
Achonry **20** A4
Aclare **20** A4
Acton **22** C2
Adamstown *South Dublin* **29** E4
Adamstown *Wexford* **39** D2
Adare **36** C2
Adrigole **41** D3
Aghaboe **32** A3
Aghabullogue **42** A2
Aghacashel **21** D4
Aghada **42** C2
Aghadiffin **26** B1
Aghadoe **41** D1
Aghadowey **16** B2
Aghadowry **28** A1
Aghagallon **22** C1
Aghagower **19** E4
Aghalane **21** E4
Aghalee **22** C1
Aghamore *Leitrim* **27** E1
Aghamore *Mayo* **26** B1
Aghamuck **27** D2
Aghanloo **16** A2
Aghavannagh **33** E3
Aghaville **41** E3
Agher **29** D3
Aghern **42** C1
Aghleam (An Eachléim) **18** C2
Aghnacliff **28** A1
Aghnagar Bridge **40** B2
Aghnamullen **22** A4
Aglish *Tipperary* **31** E3
Aglish *Waterford* **43** D1
Ahafona **35** D2
Ahakista **41** D4
Ahascragh **26** C4
Ahenny **38** B2
Aherla **42** A2
Ahoghill **16** C3
Ailt an Chorráin
  (Burtonport) **14** B3
Aldergrove **22** C1
Allen **28** C4
Allenwood **28** C4
Allihies **40** B3
Alloon **26** C4
Altagowlan **20** C4
Altbrean **21** D3
Altishane **15** F4
Altnapaste **15** D2
An Aird (Ard) **24** C3
An Aird Mhóir (Ardmore) **25** D3
An Baile Breac (Ballybrack) **40** B2
An Baile Nua (Newtown) **36** C4
An Baile Thiar (West Town) **14** C1
An Bhlarna (Blarney) **42** B2
An Bun Beag (Bunbeg) **14** C2
An Cabhán (Cavan) **21** F4
An Caiseal (Cashel) *Galway* **25** D2
An Caiseal (Cashel) *Mayo* **18** C3
An Caisleán Nua (Newcastle West)
  *Limerick* **36** B3
An Caisleán Nua (Newcastle)
  *South Dublin* **29** D4
An Caisleánn Riabhach
  (Castlerea) **26** C2
An Carn Mór (Carnmore) **26** A4
An Cathair (Cahir) **37** F3
An Charraig (Carrick) **20** A1
An Cheathrú Rua (Carraroe) **25** D3
An Chill (Kill) **30** B3
An Chillín Liath
  (Killeenleagh) **40** B2
An Chloich Mhóir
  (Cloughmore) **18** C4
An Chorrchloch (Corclogh) **18** C2
An Cionn Garbh (Kingarrow) **14** C3
An Clochán (Cloghan)
  *Donegal* **15** D4
An Clochán (Cloghane)
  *Kerry* **34** B4
An Clochán Liath (Dungloe) **14** B3
An Cóbh (Cobh) **42** B2
An Coimín (Commeen) **15** D4
An Coireán (Waterville) **40** B2
An Corrán (Corraun) **18** C3
An Daingean (Dingle) **34** B4
An Dúchoraidh (Doochary) **14** C3
An Dumhach Bheag
  (Dooghbeg) **19** D4
An Eachléim (Aghleam) **18** C2
An Fál Carrach (Falcarragh) **14** C2
An Fál Mór (Fallmore) **18** C3
An Féar Bán (Ferbane) **27** E4
An Fhairche (Clonbur) **25** E2
An Fheothanach
  (Feohanagh) **34** A4
An Fód Dubh (Blacksod) **18** C3
An Geata Mór
  (Binghamstown) **18** C2
An Ghrafaidh (Graffy) **14** C4
An Goirtín (Gorteen) **38** A4
An Gort (Gort) **30** C2

An Gort Mór (Gortmore) **25** D3
An Leadhb Gharbh
  (Leabgarrow) **14** B3
An Longfort (Longford) **27** E2
An Machaire (Maghera)
  *Donegal* **14** B4
An Machaire (Maghery)
  *Donegal* **14** B4
An Mám (Maum) **25** D2
An Mhala Raithní (Mulrany) **19** D4
An Mhuiríoch (Murreagh) **34** A4
An Móta (Moate) **27** E4
An Muileann gCearr
  (Mullingar) **28** B3
An Nás (Naas) **33** D2
An Pasáiste (Passage West) **42** B2
An Ráth (Charleville) **36** C3
An Rinn (Ringville) **38** A4
An Ros (Rush) **29** F3
An Sciobairín (Skibbereen) **41** E4
An Spidéal (Spiddal) **25** E3
An Sráidbhaile (Stradbally) **34** B4
An tAonach (Nenagh) **31** E4
An Teach Dóite
  (Maam Cross) **25** D2
An Teampall Mór
  (Templemore) **31** F4
An Tearmann (Termon) **15** D2
An tInbhear (Inver) **18** C2
An tInbhear Mór (Arklow) **33** F4
An Trian (Trean) **25** E1
An Trianlarach
  (Teeranearagh) **40** B2
An tSeanga Mheáin
  (Tangavane) **14** C4
An tSraidbhaile (Stradbally) **32** B3
An tSraith (Srah) **25** E1
An Tulch (Tullow) **33** D4
An Uaimh (Navan) **29** D2
Anagaire (Annagry) **14** B3
Anascaul **34** B4
Anglesborough **37** E3
Anlore **21** F3
Annacarriga **31** D4
Annacarty **37** D2
Annacloghmullin **22** B3
Annaclone **22** C2
Annacloy **23** E2
Annacotty **37** D2
Annacurragh **33** E4
Annadorn **23** E2
Annagassan **29** E1
Annagh **20** A4
Annaghdown
  (Eanach Dhúin) **26** A4
Annaghmore **27** E1
Annagry (Anagaire) **14** B3
Annahilt **23** D2
Annalong **23** D4
Annamoe **33** E3
Annayalla **22** A3
Annestown **38** B4
Annfield **31** F4
Annsborough **23** D3
Antrim **16** C4
Araglin **37** E4
Archerstown **28** B2
Ard (An Aird) **24** C3
Ard an Rátha (Ardara) **14** B4
Ard na Caithne (Smerwick) **34** A4
Ardabrone **20** A3
Ardagh *Limerick* **36** B3
Ardagh *Longford* **28** A2
Ardan **28** A4
Ardanairy **33** F3
Ardara (Ard an Rátha) **14** B4
Ardarragh **22** C3
Ardattin **33** D4
Ardboe **22** B1
Ardcath **29** E2
Ardcroney **31** E3
Ardea **40** C2
Ardee (Baile Átha Fhirdhia) **29** D1
Ardfert **34** C3
Ardfield **41** F4
Ardfinnan **37** F4
Ardgehane **42** A4
Ardglass *Cork* **42** C1
Ardglass
  *Newry, Mourne & Down* **23** E3
Ardgroom **40** C3
Ardkeen **23** F2
Ardlougher **21** E4
Ardmillan **23** E1
Ardmore (An Aird Mhóir)
  *Galway* **25** D3
Ardmore *Roscommon* **20** B4
Ardmore *Waterford* **43** D2
Ardmorney **28** A4
Ardnacrusha **30** C4
Ardnagreevagh **24** C3
Ardnasodan **26** B3
Ardpatrick **37** D3
Ardrahan **30** C2
Ardscull **32** C3
Ardstraw **15** E4
Arghavas **21** D4
Arigna **20** C4
Arklow (An tInbhear Mór) **33** F4
Arless **32** C3
Armagh **22** B2
Armoy **16** C2
Arney **21** E3
Arryheernabin
  (Earra Thíre na Binne) **15** E1
Arthurstown **38** C3
Articlave **16** A2
Artigarvan **15** E3
Artikelly **16** A2

Arvagh **28** A1
Ashbourne (Cill Dhéagláin) **29** E3
Ashford *Limerick* **36** B3
Ashford *Wicklow* **33** F3
Askamore **39** E1
Askanagap **33** E3
Askeaton **36** B2
Astee **35** D2
Áth an Sceire (Enniskerry) **33** F2
Áth Trasna (Newmarket) **36** B4
Athboy (Baile Átha Buí) **28** C2
Athea **35** E3
Athenry (Baile Átha an Rí) **26** B4
Athgarvan **33** D2
Athlacca **36** C3
Athleague **37** D3
Athlone (Baile Átha Luain) **27** E3
Athlumney **29** D2
Athy (Baile Átha Í) **32** C3
Attanagh **32** B4
Attavalley **19** D2
Attical **23** D4
Attymass **19** F3
Attymon **26** B4
Aughacasla **34** C4
Aughagault **15** D3
Aughamullan **22** B1
Augher **21** F2
Aughils Bridge **34** C4
Aughnacloy **22** A2
Aughnasheelan **21** D4
Aughnaskeagh **23** D2
Aughrim *Clare* **30** B3
Aughrim *Galway* **26** C4
Aughrim *Mid Ulster* **16** B4
Aughrim *Wicklow* **33** E4
Aughris **20** A3
Aughrus More **24** B2
Avoca **33** F3

## B

Bagenalstown
  (Muine Bheag) **32** C4
Baile an Bhiataigh
  (Bettystown) **29** E2
Baile an Bhuinneánaigh
  (Ballybunion) **35** D2
Baile an Chinnéidigh (Newtown
  Mount Kennedy) **33** F2
Baile an Dúlaigh
  (Ballindooly) **26** A4
Baile an Fheirtéaraigh
  (Ballyferriter) **34** A4
Baile an Lochaigh
  (Ballinloghig) **34** B4
Baile an Mhóta (Ballymote) **20** B4
Baile an Mhuilinn (Milltown) **34** B4
Baile an Róba (Ballinrobe) **25** E1
Baile an Sceilg
  (Ballinskelligs) **40** B2
Baile Átha an Rí (Athenry) **26** B4
Baile Átha Buí (Athboy) **28** C2
Baile Átha Cliath (Dublin) **29** F3
Baile Átha Fhirdhia (Ardee) **29** D1
Baile Átha Í (Athy) **32** C3
Baile Átha Luain (Athlone) **27** E3
Baile Átha Troim (Trim) **29** D3
Baile Bhuirne (Ballyvourney) **41** E2
Baile Brigín (Balbriggan) **29** F2
Baile Chláir (Claregalway) **26** A4
Baile Coimin (Blessington) **33** D2
Baile Ghib (Gibstown) **29** D2
Baile Loch Riach (Loughrea) **31** D2
Baile Mhic Andáin
  (Thomastown) **38** B1
Baile Mhic Íre
  (Ballymakeery) **41** F2
Baile Mhistéale
  (Mitchelstown) **37** E4
Baile na Finne (Fintown) **14** C3
Baile na Lorgan
  (Castleblayney) **22** B4
Baile Órthaí (Oristown) **28** C2
Baile Uí Fhuaruisce
  (Ballyhoorisky) **15** D1
Baile Uilcín (Wilkinstown) **29** D2
Bailieborough
  (Coillan Chollaigh) **28** C1
Balbriggan (Baile Brigín) **29** F2
Baldwinstown **29** E3
Balla **26** A1
Ballagh *Galway* **26** B2
Ballagh *Limerick* **36** B3
Ballagh *Roscommon* **27** D3
Ballagh *Roscommon* **27** D2
Ballaghaderreen
  (Bealach A Doirín) **26** C1
Ballaghbehy **35** E3
Ballaghkeen **39** E2
Ballaghmore **32** A3
Ballaghnatrillack **20** B2
Ballard **33** D4
Ballardiggan **30** C2
Balleen **32** A4
Balleer **22** B3
Ballickmoyler **32** C3
Balliggan **23** F1
Ballina (Béal an Átha)
  *Mayo* **19** F3
Ballina *Tipperary* **31** D4
Ballinaboy **24** C2
Ballinabrackey **28** B3
Ballinabranagh **32** C4
Ballinacarrow **20** B3
Ballinaclash **23** E1
Ballinacor **33** F3
Ballinadee **42** A3
Ballinafad **20** B4

Ballinagar **28** B4
Ballinagh **28** A1
Ballinagore **28** A4
Ballinagree **41** F1
Ballinakill **28** B3
Ballinalack **28** A2
Ballinalee **28** A1
Ballinamallard **21** E3
Ballinameen **27** D1
Ballinamore **21** D4
Ballinamuck **27** E1
Ballinamult **37** F4
Ballinascarthy **42** A3
Ballinasloe
  (Béal Átha na Sluaighe) **27** D4
Ballinclashet **42** B3
Ballinclogher **35** D3
Ballincollig **42** A2
Ballincrea **38** C3
Ballincurrig **42** C1
Ballindaggan **39** D1
Ballindarragh **21** E3
Ballinderreen **30** B2
Ballinderry *Tipperary* **31** E3
Ballinderry *Wicklow* **33** E3
Ballindine **26** A2
Ballindooly
  (Baile an Dúlaigh) **26** A4
Ballindrait **15** E4
Ballineen **41** F3
Ballinfull **20** B2
Ballingarry *Limerick* **36** C3
Ballingarry *Tipperary* **31** E3
Ballingarry *Tipperary* **38** A1
Ballingeary (Béal Átha an
  Ghaorthaidh) **41** F2
Ballingurteen **41** F3
Ballinhassig **42** B2
Ballinleeny **36** C3
Ballinlig **28** C2
Ballinloghig
  (Baile an Lochaigh) **34** B4
Ballinlough *Meath* **28** C2
Ballinlough *Roscommon* **26** B2
Ballinrobe (Baile an Róba) **25** E1
Ballinruan **30** C3
Ballinskelligs
  (Baile an Sceilg) **40** B2
Ballinspittle **42** A3
Ballinteer **16** B2
Ballintober **26** C2
Ballintogher **20** B3
Ballintoy **16** C1
Ballintra **20** C1
Ballintubber **25** E1
Ballintubbert **32** C3
Ballinure **37** F2
Ballinvarry **38** C2
Ballinwillin **42** C1
Ballitore **32** C3
Ballivor **28** C3
Ballon **33** D4
Balloor **20** B2
Balloughter **39** E1
Ballure *Donegal* **20** C1
Ballure *Sligo* **20** C3
Ballyagran **36** C3
Ballyalton **23** E2
Ballyard **21** E2
Ballybarrack **22** C4
Ballybay (Béal Átha Beithe) **22** A3
Ballybofey (Bealach Feich) **15** D4
Ballyboghil **29** E3
Ballybogy **16** B2
Ballybornia **27** E3
Ballyboy **22** A2
Ballybrack (An Baile Breac) **40** B2
Ballybriest **16** A4
Ballybrittas **32** B3
Ballybrophy **32** A3
Ballybunion (Baile an
  Bhuinneánaigh) **35** D2
Ballycahill **37** F2
Ballycallan **38** B1
Ballycanew **39** E1
Ballycar **30** C4
Ballycarney **39** D1
Ballycarry **17** E4
Ballycassidy **21** E2
Ballycastle *Mayo* **19** E2
Ballycastle
  *Causeway Coast & Glens* **16** C1
Ballyclare **17** D4
Ballyclerahan **37** F3
Ballyclough **36** C4
Ballycogly **39** E3
Ballycolla **32** A3
Ballyconneely **24** C2
Ballyconnell *Cavan* **21** E4
Ballyconnell *Sligo* **20** A2
Ballycorkey **28** A2
Ballycotton **42** C2
Ballycraigy **17** D3
Ballycrossaun **31** E2
Ballycroy **19** D3
Ballycullane **38** C2
Ballycumber **28** A4
Ballydan **23** E2
Ballydangan **27** D4
Ballydavid **31** D2
Ballydavis **32** B2
Ballydehob **41** D4
Ballydonegan **40** B3
Ballydoolagh **21** E2
Ballydrain **23** E1
Ballyduff *Kerry* **35** D3
Ballyduff *Waterford* **42** C1
Ballyduff *Wexford* **39** E1

Ballydugan **23** E2
Ballyea **36** B3
Ballyeaston **17** D4
Ballyesborough **23** F1
Ballyfad **33** E4
Ballyfarnagh **26** A1
Ballyfarnon **20** C4
Ballyfasy **38** C2
Ballyfeard **42** B3
Ballyferriter
  (Baile an Fheirtéaraigh) **34** A4
Ballyfin **32** A2
Ballyforan **27** D3
Ballyfore **28** B4
Ballyfoyle **32** B4
Ballygalley **17** D3
Ballygar **26** C3
Ballygarrett **39** F1
Ballygarries **26** A2
Ballygawley *Mid Ulster* **22** A2
Ballygawley *Sligo* **20** B3
Ballyglass **26** A2
Ballygorian **22** C3
Ballygorman **15** F1
Ballygowan
  *Ards & North Down* **23** E1
Ballygowan
  *Newry, Mourne & Down* **23** D4
Ballyhack **38** C3
Ballyhaght **36** C4
Ballyhahill **35** E2
Ballyhaise **21** F4
Ballyhalbert **23** F1
Ballyhale *Galway* **26** A3
Ballyhale *Kilkenny* **38** B2
Ballyhar **41** D1
Ballyhaunis
  (Béal Átha hAmhnais) **26** B2
Ballyhea **36** C4
Ballyhean **19** E4
Ballyhear **26** A3
Ballyheerin **15** D2
Ballyheigue **34** C3
Ballyhoe Bridge **16** C2
Ballyhoge **39** D2
Ballyholme **17** E4
Ballyhooly **42** B1
Ballyhoorisky
  (Baile Uí Fhuaruisce) **15** D1
Ballyhornan **23** E2
Ballyhosset **23** E2
Ballyhugh **21** E4
Ballyjamesduff **28** B1
Ballykean **32** B2
Ballykeeran **27** E3
Ballykelly **16** A2
Ballykilleen **28** C4
Ballykinler **23** E3
Ballylaghnan **31** D3
Ballylanders **37** D3
Ballylaneen **38** B3
Ballylawn **15** E3
Ballyleague **27** E2
Ballyleny **22** B2
Ballylickey **41** E3
Ballyliffin **15** E1
Ballyline **38** A1
Ballylongford **35** D2
Ballylooby **37** F3
Ballyloughbeg **16** B2
Ballylucas **21** E2
Ballylynan **32** C3
Ballymacarberry **37** F4
Ballymacashen **23** E1
Ballymacavany **21** D1
Ballymacaw **38** C4
Ballymacelligott **35** D4
Ballymachugh **28** B1
Ballymack **38** B1
Ballymackey **31** E4
Ballymackilroy **22** A2
Ballymacmague **38** A4
Ballymacoda **43** D2
Ballymaconnelly **16** B3
Ballymacran **16** A2
Ballymacurley **27** D2
Ballymacward **26** C4
Ballymadog **43** D2
Ballymagaraghy **16** A1
Ballymagorry **15** E3
Ballymahon **27** E3
Ballymakane **28** C1
Ballymakeery
  (Baile Mhic Íre) **41** F2
Ballymakenny **29** E1
Ballymanus **28** B3
Ballymartin **23** D4
Ballymartle **42** B3
Ballymascanlan **22** C4
Ballymena **16** C3
Ballymoe **26** C2
Ballymore *Donegal* **15** D2
Ballymore *Westmeath* **28** A3
Ballymore Eustace **33** D2
Ballymorris **38** B3
Ballymote
  (Baile an Mhóta) **20** B4
Ballymurn **39** E2
Ballymurphy **38** C1
Ballymurry **27** D2
Ballynabola **38** C2
Ballynacally **30** B4
Ballynacarriga **41** F3
Ballynacarragh **28** A3
Ballynaclogh **31** E4
Ballynacorra **42** C2
Ballynacourty **38** A4

Ballynadrumny **28** C3
Ballynahatty **21** F1
Ballynahinch
  *Newry, Mourne & Down* **23** D2
Ballynahinch *Galway* **24** C2
Ballynahowen **27** E4
Ballynakill *Carlow* **38** C1
Ballynakill *Westmeath* **27** E3
Ballynakilla **40** C3
Ballynakilly *Mid Ulster* **22** B1
Ballynakilly *Kerry* **40** C3
Ballynamallaght **15** F4
Ballynamona *Cavan* **28** C1
Ballynamona *Cork* **42** A1
Ballynamona *Galway* **26** B3
Ballynamore Bridge **26** C3
Ballynare **29** D3
Ballynarry **28** B1
Ballynaskreena **35** D3
Ballyneaner **15** F3
Ballyneety **37** D2
Ballyneill **38** A2
Ballynockan **33** E2
Ballynoe *Cork* **42** C1
Ballynoe
  *Newry, Mourne & Down* **23** E3
Ballynora **42** A2
Ballynunty **38** A1
Ballynure **17** D4
Ballyorgan **37** D4
Ballypatrick **38** A2
Ballyporeen **37** E4
Ballyragget **32** B4
Ballyrashane **16** B2
Ballyreagh **22** A2
Ballyroan **32** B3
Ballyroebuck **39** D1
Ballyrogan **16** B3
Ballyronan **16** B4
Ballyroney **23** D3
Ballyroon **40** C4
Ballysadare **20** B3
Ballyshannon (Béal Átha Seanaidh)
  *Donegal* **20** C1
Ballyshannon *Kildare* **32** C2
Ballysheedy **36** C2
Ballysimon **39** E1
Ballysloe **38** A1
Ballysteen **36** B2
Ballystrudder **17** E4
Ballytoohy **18** C4
Ballyvaldon **39** E2
Ballyvaughan **30** B2
Ballyveagh **23** D4
Ballyvoneen **26** C3
Ballyvourney
  (Baile Bhuirne) **41** E2
Ballyvoy **16** C1
Ballyvoyle **38** A4
Ballywalter **23** F1
Ballyward **23** D3
Ballywatticock **23** E1
Ballywilliam **38** C2
Balnamore **16** B2
Balrath **29** E2
Balrothery **29** E2
Balscaddan **29** E2
Baltimore **41** E4
Baltinglass (Bealach Conglais) **33** D3
Baltray **29** F2
Banada **20** A4
Banagher (Beannchar) **31** F2
Banbridge **22** C2
Bandon
  (Droichead na Bandan) **42** A3
Banemore **35** D3
Bangor *Mayo* **19** D2
Bangor *Ards & North Down* **17** E4
Banna **34** C3
Bannfoot **22** B1
Bannow **39** D3
Banoge **38** C3
Bansha **37** E3
Banteer **41** F1
Bantry (Beantraí) **41** D3
Barefield **30** B3
Barna (Bearna) *Galway* **26** A4
Barna *Limerick* **37** E3
Barna *Offaly* **31** F3
Barnacahogue **26** B1
Barnaderg **26** B3
Barnagh **36** B2
Barnatra (Barr na Trá) **18** C2
Barneycarroll **26** A2
Barr na Trá (Barnatra) **18** C2
Barraduff **41** E1
Barrigone **36** B2
Barringtonsbridge **37** D2
Barry **27** E2
Batterstown **29** D3
Baunskeha **38** B2
Bauntlieve **30** A3
Bawn Cross Roads **36** B4
Bawnard **42** C2
Bawnboy **21** E4
Baylet **15** E2
Beagh **26** B3
Beal **35** D2
Béal an Átha (Ballina) **19** F3
Béal an Átha Móir
  (Bellanamore) **14** C3
Béal an Mhuirthead
  (Belmullet) **18** C2
Béal Átha an Ghaorthaidh
  (Ballingeary) **41** E2
Béal Átha Beithe
  (Ballybay) **22** A3

Béal Átha hAmhnais
  (Ballyhaunis) **26** B2
Béal Átha Liag
  (Lanesborough) **27** E2
Béal Átha na Muice
  (Swinford) **26** A1
Béal Átha na Sluaighe
  (Ballinasloe) **27** D4
Béal Átha Seanaidh
  (Ballyshannon) **20** C1
Béal Deirg (Belderg) **19** D2
Béal Easa (Foxford) **19** F3
Bealach A Doirín
  (Ballaghaderreen) **26** C1
Bealach Conglais
  (Baltinglass) **33** D3
Bealach Feich (Ballybofey) **15** D4
Bealaclugga **30** B2
Bealad **41** F3
Bealaha **35** D1
Bealalaw Bridge **40** C1
Bealnablath **42** A2
Beannchar (Banagher) **31** F2
Beantraí (Bantry) **41** D3
Bearna (Barna) **26** A4
Beaufort **41** D1
Bective **29** D3
Bekan **26** B2
Belcarra **19** E4
Belclare **26** A3
Belcoo **21** D3
Belderg (Béal Deirg) **19** D2
Belfast **23** D1
Belfast International Airport **16** C4
Belgooly **42** B3
Bellacorick **19** D2
Bellaghy *Mid Ulster* **16** B4
Bellaghy *Sligo* **20** A4
Bellanagare **26** C1
Bellanaleck **21** E3
Bellanamore
  (Béal an Átha Móir) **14** C3
Bellanamullia **27** D3
Bellaneeny **27** D4
Bellanode **22** A3
Bellavary **26** A1
Belleek *Fermanagh & Omagh* **20** C2
Belleek
  *Newry, Mourne & Down* **22** B3
Bellewstown **29** E2
Bellview **28** B3
Belmont **17** E4
Belmount **27** E4
Belmullet
  (Béal an Mhuirthead) **18** C2
Beltra **19** E3
Belturbet **21** E4
Belville **19** E2
Benburb **22** B2
Bendooragh **16** B2
Bennekerry **32** C4
Bennettsbridge **38** B1
Beragh **21** F1
Berrings **42** A2
Bessbrook **22** C3
Bettystown
  (Baile an Bhiataigh) **29** E2
Bilboa **32** C4
Binghamstown (An Geata Mór) **18** C2
Biorra (Birr) **31** F2
Birdhill **31** D4
Birr (Biorra) **31** F2
Bishops Court **23** E2
Black Hills **29** F2
Black Lion **32** A2
Black Skull **22** C2
Blackcastle **29** D2
Blackfort **21** F1
Blacklion **21** D3
Blackrock
  *Dún Laoghaire-Rathdown* **29** F4
Blackrock *Louth* **22** C4
Blacksod (An Fód Dubh) **18** C3
Blacktown **21** E1
Blackwater **39** E2
Blackwater Bridge **41** D2
Blackwatertown **22** B2
Blanchardstown **29** E4
Blaney **21** D2
Blarney (An Bhlarna) **42** B2
Bleary **22** C2
Blennerville **35** D4
Blessington
  (Baile Coimin) **33** D2
Blue Ball **32** A2
Blueford **36** B4
Boardmills **23** D2
Bodoney **21** E1
Bodyke **30** C3
Bofeenaun **19** E3
Bohaun **25** E1
Boher **37** D2
Boheraphuca **31** F3
Boherbue **36** B4
Boherlahan **37** F2
Bohermeen **29** D2
Boho **21** D2
Bohola **26** A1
Bolea **16** A2
Boleran **16** B3
Boley **32** C2
Boola **43** D1
Boolavogue **39** E1
Boolteens **35** D4
Boolyglass **38** B2
Borris **38** C1
Borris-in-Ossory **32** A3
Borrisokane **31** E2
Borrisoleigh **31** F4
Boston **30** B3

Bottlehill **42** B1
Boulduff **31** F4
Bovedy **16** B3
Boviel **16** A3
Boyerstown **29** D2
Boyle (Mainistir na Búille) **20** C4
Brackagh **28** B4
Brackary **20** C2
Brackloon **26** B1
Bracklyn **28** B3
Bracknagh **32** C2
Braddocks **22** A3
Brandon (Cé Bhréanainn) **34** B4
Brannockstown **33** D2
Bray (Bré) **33** F2
Bré (Bray) **33** F2
Bready **15** E3
Breaghva **35** D2
Bree **39** D2
Breenagh (Na Bruíneacha) **15** D3
Brickens **26** B2
Bridebridge **42** C1
Brideswell **27** D3
Bridge End **15** E2
Bridgeland **33** E4
Bridgetown *Donegal* **20** C1
Bridgetown *Wexford* **39** D3
Brinlack (Bun na Leaca) **14** C2
Brittas *Limerick* **37** D2
Brittas *South Dublin* **29** D4
Brittas Bridge **33** F3
Britway **42** C1
Broadford *Clare* **30** C4
Broadford *Limerick* **36** B3
Broadway **39** E3
Brookeborough **21** E2
Brooklodge **42** B2
Broomfield **22** B4
Brosna *Kerry* **35** E4
Brosna *Offaly* **31** F3
Broughal **31** F2
Broughderg **16** A4
Broughshane **16** C3
Brownsgrove **26** B3
Brownstown **38** C4
Bruckless **20** B1
Bruff **37** D3
Bruree **36** C3
Bryansford **23** D3
Buckna **17** D3
Buckode **20** C2
Buggan **21** D2
Bulgaden **37** D3
Bullaun **26** C4
Bun an Churraigh
  (Bunacurry) **18** C3
Bun an Phobail (Moville) **16** A2
Bun an Tábhairne
  (Crosshaven) **42** B2
Bun Clóidí (Bunclody) **39** D1
Bun Cranncha (Buncrana) **15** E2
Bun Dobhrain (Bundoran) **20** C2
Bun na hAbhna
  (Bunnahowen) **18** C2
Bun na Leaca (Brinlack) **14** C2
Bunacurry
  (Bun an Churraigh) **18** C3
Bunaw **40** C2
Bunbeg (An Bun Beag) **14** C2
Bunclody (Bun Clóidí) **39** D1
Buncrana (Bun Cranncha) **15** E2
Bundoran (Bun Dobhrain) **20** C2
Bunglass **28** C3
Bunlahy **28** B1
Bunmahon **38** B4
Bunnaglass **30** C2
Bunnahowen
  (Bun na hAbhna) **18** C2
Bunnanaddan **20** B4
Bunneil **21** F3
Bunnyconnellan **19** F3
Burnchurch **38** B1
Burncourt **37** E4
Burnfoot *Donegal* **15** E2
Burnfoot
  *Causeway Coast & Glens* **16** A3
Burnfort **42** A1
Burnside
  *Antrim & Newtownabbey* **17** D4
Burnside
  *Antrim & Newtownabbey* **16** C4
Burren *Clare* **30** B2
Burren *Mayo* **19** E4
Burren
  *Newry, Mourne & Down* **22** C3
Burrow **39** E3
Burtonport
  (Ailt an Chorráin) **14** B3
Bushmills **16** B1
Butlers Bridge **21** F4
Butlerstown **42** A4
Butterlope **15** F4
Buttevant (Cill na Mallach) **36** C4
Bweeng **42** A1

## C

Cabragh **22** A1
Cadamstown *Kildare* **28** C4
Cadamstown *Offaly* **32** A2
Caddy **16** C4
Caher **30** C4
Caheragh **41** E3
Caherbarnagh **41** E1
Caherconlish **37** D2
Caherea **30** B4
Caherline **37** D2
Cahermore *Cork* **40** B3
Cahermore *Galway* **30** C2

Cahernarry **37** D2
Cahir (An Cathair) **37** F3
Cahirsiveen
  (Cathair Saidhbhín) **40** B2
Caiseal (Cashel) **37** F2
Caisleán an Bharraigh
  (Castlebar) **19** E4
Caisleán an Chomair
  (Castlecomer) **32** B4
Caisleán Uí Chonail
  (Castleconnell) **31** D4
Caldanagh **16** C2
Caledon **22** A2
Calhame **20** B1
Callan (Callain) **38** B1
Callan (Callain) **38** B1
Callow **24** C2
Caltra **26** C3
Caltraghlea **26** C4
Calverstown **33** D2
Camaross **39** D2
Camlough **22** C3
Camolin **39** E1
Camp **34** C4
Campile **38** C3
Campsey **15** F2
Camross **32** A3
Canningstown **22** A4
Cappagh *Armagh City,
  Banbridge & Craigavon* **22** C2
Cappagh *Mid Ulster* **22** A1
Cappagh *Limerick* **36** B2
Cappagh White **37** E2
Cappaghmore **30** B2
Cappalinnan **32** A4
Cappamore **37** D2
Cappataggle **26** C4
Cappeen **41** F2
Cappoquin **43** D1
Capragh **22** B4
Caraunkeelwy **26** A4
Carbury **28** C4
Cardy **23** E1
Cargan **16** C3
Carland **22** A1
Carlanstown **28** C2
Carlingford **22** C4
Carlow (Ceatharlach) **32** C4
Carna **24** C3
Carnanreagh **15** F3
Carnaross **28** C2
Carncastle **17** D3
Carndonagh **15** F1
Carnduff **16** C1
Carnew **33** E4
Carney **20** B2
Carnlough **17** D3
Carnmore (An Carn Mór) **26** A4
Carnowen **15** E4
Carnteel **22** A2
Carracastle **20** A4
Carragh **29** D4
Carraholly **19** D4
Carraig Airt (Carrickart) **15** D2
Carraig na Siuire
  (Carrick-on-Suir) **38** A2
Carraig Thuathail (Carrigtohill) **42** C2
Carraig Uí Leighin (Carrigaline) **42** B2
Carran **30** B3
Carraroe (An Cheathru Rua) **25** D3
Carras **26** A2
Carrick (An Charraig) *Donegal* **20** A1
Carrick *Fermanagh & Omagh* **21** E1
Carrick *Wexford* **39** D3
Carrickaboy **28** B1
Carrickart (Carraig Airt) **15** D2
Carrickashedoge **29** D1
Carrickbeg **38** A2
Carrickboy **28** A2
Carrickfergus **17** E4
Carrickmacross
  (Carrig Mhachaire) **22** B4
Carrickmore **22** A2
Carricknagower **27** E3
Carrick-on-Shannon
  (Cora Droma Rúisc) **27** D1
Carrick-on-Suir
  (Carraig na Siuire) **38** A2
Carrickroe **22** A2
Carrig *Cork* **37** D4
Carrig *Tipperary* **31** F2
Carrig Mhachaire
  (Carrickmacross) **22** B4
Carrigagulla **41** F1
Carrigaholt **35** D2
Carrigahorig **31** E2
Carrigaline
  (Carraig Uí Leighin) **42** B2
Carrigallen **21** E4
Carrigan **28** A1
Carriganima **41** F1
Carrigans **15** E3
Carrigatogher **31** E2
Carrigeenroe **20** C4
Carrigenagh **23** D4
Carrigkerry **36** B3
Carrignavar **42** B1
Carrigtohill
  (Carraig Thuathail) **42** C2
Carrowkeel *Donegal* **15** E2
Carrowkeel *Donegal* **15** F2
Carrowkeel *Galway* **31** D2
Carrowkennedy **25** D1
Carrowmenagh **15** F1
Carrowmore *Galway* **30** B2
Carrowmore *Sligo* **20** A4
Carrowmore Lacken **19** E2
Carrowmoreknock **26** A3
Carrownacon **25** E1
Carrowneden **20** A3

Carrownisky **24** C1
Carrowntreila **19** F3
Carrownurlar **26** C1
Carrowreagh *Roscommon* **27** D4
Carrowreagh *Sligo* **20** A3
Carrowteige (Ceathrú Thaidhg) **19** D1
Carryduff **23** D1
Cashel *Cavan* **28** B1
Cashel *Donegal* **15** D2
Cashel *Galway* **26** B2
Cashel (An Caiseal) *Galway* **25** D2
Cashel *Laois* **32** B3
Cashel (An Caiseal) *Mayo* **18** C3
Cashel (Caiseal) *Tipperary* **37** F2
Cashelgarran **20** B2
Cashla **26** B4
Cashleen **24** C1
Cashty **15** F4
Casla (Costelloe) **25** D3
Castlebaldwin **20** B4
Castlebar
  (Caisleán an Bharraigh) **19** E4
Castlebellingham **29** E1
Castleblakeney **26** C3
Castleblayney
  (Baile na Lorgan) **22** B4
Castlebridge **39** E2
Castlecaulfield **22** A1
Castlecomer
  (Caisleán an Chomair) **32** B4
Castleconnell
  (Caisleán Uí Chonail) **31** D4
Castleconor **19** F2
Castlecor **36** C4
Castlecove **40** B2
Castlecuffe **32** A2
Castledawson **16** B4
Castlederg **15** F4
Castledermot **32** C3
Castledonovan **41** E3
Castlefinn **15** E4
Castlefreke **41** F4
Castlegal **20** B2
Castlegregory **34** C4
Castlehill *Mayo* **19** D3
Castlehill *Mayo* **19** E3
Castleisland
  (Oileán Ciarraí) **35** E4
Castlejordan **28** B4
Castleknock **29** E4
Castleleiny **31** F4
Castlellis **39** E2
Castlelyons **42** C1
Castlemaine **35** D4
Castlemartyr **42** C2
Castleplunket **26** C2
Castlepollard **28** B2
Castlerahan **28** B1
Castlerea
  (An Caisleánn Riabhach) **26** C2
Castlereagh **23** D1
Castlerickard **28** C3
Castlerock **16** A2
Castleroe **16** B2
Castleruddery **33** D3
Castlesampson **27** D3
Castleshane **22** A3
Castletown *Clare* **30** B3
Castletown *Cork* **41** F3
Castletown *Laois* **32** A3
Castletown *Limerick* **36** C3
Castletown *Louth* **22** C4
Castletown *Meath* **29** D1
Castletown *Westmeath* **28** B2
Castletown *Wexford* **33** F4
Castletown Geoghegan **28** A3
Castletownbere **40** C3
Castletownroche **37** D4
Castletownshend **41** E4
Castlewarren **32** C4
Castlewellan **23** D3
Cathair Dónall (Catherdaniel) **40** B3
Cathair na Mart (Westport) **19** D4
Cathair Saidhbhín
  (Cahirsiveen) **40** B2
Catherdaniel (Cathair Dónall) **40** B3
Causeway **35** D3
Causeway Head **16** B1
Cavan (An Cabhán) **21** F4
Cavanagarven **22** A3
Cavangarden **20** C1
Cé Bhréanainn (Brandon) **34** B4
Ceanannus Mór (Kells) **28** C2
Ceann Toirc (Kanturk) **36** B4
Ceann Trá (Ventry) **34** A4
Ceatharlach (Carlow) **32** C4
Ceathrú Thaidhg
  (Carrowteige) **19** D1
Cecilstown **36** C4
Celbridge (Cill Droichid) **29** D4
Chapeltown
  *Mid & East Antrim* **16** C4
Chapeltown
  *Newry, Mourne & Down* **23** E2
Chapeltown *Kerry* **34** A2
Chapeltown *Kerry* **40** A2
Charlemont **22** B2
Charlestown **20** A4
Charleville (An Ráth) **36** C3
Cheekpoint **38** C3
Cherryville **32** C4
Church Ballee **23** E2
Church Cross **41** E4
Church Hill *Donegal* **15** D3
Church Hill
  *Fermanagh & Omagh* **21** D2
Church Town **15** E3
Churchtown *Mid Ulster* **16** B4
Churchtown *Cork* **36** C4
Churchtown *Cork* **42** C2

Churchtown
  Newry, Mourne & Down 23 E2
Churchtown Kildare 32 C3
Churchtown Limerick 36 B3
Churchtown Wexford 39 E3
Churchtown Wexford 38 C4
Cill Airne (Killarney) 41 D1
Cill Bhríde (Kilbride) 28 C2
Cill Chainnigh (Kilkenny) 38 B1
Cill Chaoi (Kilkee) 35 D2
Cill Charthaigh (Kilcar) 20 B1
Cill Chiaráin (Kilkieran) 25 D3
Cill Choca (Kilcock) 29 D4
Cill Chomhghaill (Kilcoole) 33 F2
Cill Chormaic (Kilcormac) 31 F2
Cill Chuillin (Kilcullen) 33 D2
Cill Chuimín (Kilcummin) 34 B4
Cill Dalua (Killaloe) 31 D4
Cill Dara (Kildare) 32 C2
Cill Dhéagláin (Ashbourne) 29 E3
Cill Éinne (Killeany) 25 D4
Cill Maoilchéadair
  (Kilmalkedar) 34 B4
Cill Mhantáin (Wicklow) 33 F3
Cill Mocheallóg (Kilmallock) 37 D3
Cill na Mallach (Buttevant) 36 C4
Cill na Martra (Kilnamartyra) 41 D2
Cill Ón Chatha (Killonecaha) 40 A2
Cill Orglan (Killorglin) 40 C1
Cill Rois (Kilrush) 35 D2
Cill Rónáin (Kilronan) 25 D4
Cinn Mhara (Kinvarra) 25 D3
Cióirtheach (Clara) 28 A4
Cionn Caslach
  (Kincasslagh) 14 B2
Cionn tSáile (Kinsale) 42 B3
City of Derry Airport 15 F2
Clabby 21 F2
Claddagh 26 B3
Claddaghduff 24 B2
Clady Mid Ulster 16 B3
Clady Derry City & Strabane 15 E4
Clady Milltown 22 B3
Claggan 15 D2
Clanabogan 21 F1
Clane (Claonadh) 29 D4
Claonadh (Clane) 29 D4
Clár Chlainne Mhuiris
  (Claremorris) 26 A2
Clara (Cióirtheach) 28 A4
Clarahill 32 A2
Clare Armagh City,
  Banbridge & Craigavon 22 C2
Clare Armagh City,
  Banbridge & Craigavon 22 C2
Clarecastle 30 B4
Clareen 31 F2
Claregalway
  (Baile Chláir) 26 A4
Claremorris
  (Clár Chlainne Mhuiris) 26 A2
Clarina 36 C2
Clarinbridge 26 B4
Clashmore 43 D1
Claudy 15 F3
Cleady 41 D2
Cleggan 24 C2
Cleristown 39 D3
Clifden 24 C2
Clifferna 21 F4
Cliffoney 20 B2
Cloch na Rón (Roundstone) 24 C2
Clogh Mid & East Antrim 16 C3
Clogh Kilkenny 32 B4
Clogh Laois 32 A3
Clogh Wexford 39 E1
Cloghan (An Clochán)
  Donegal 15 D4
Cloghan Offaly 31 F2
Cloghan Westmeath 28 B3
Cloghane (An Clochán) 34 B4
Cloghans Hill 26 A2
Cloghaun 26 A4
Cloghboley 20 B3
Cloghboy 14 B4
Cloghbrack 28 C3
Cloghcor 15 E3
Clogheen Tipperary 37 F4
Clogheen Waterford 37 F4
Clogher Mid Ulster 21 F2
Clogher Kerry 35 D4
Clogher Mayo 25 E1
Clogher Mayo 19 E4
Clogher Roscommon 27 D1
Cloghera 30 C4
Clogherhead 29 E1
Cloghfin 15 E4
Cloghoge 22 C3
Cloghran 29 E3
Cloghroe Cork 42 A2
Cloghroe Donegal 15 D3
Cloghy 23 F2
Clohamon 39 D1
Clohernagh 38 C3
Cloich na Coillte (Clonakilty) 41 F3
Clomantagh 32 A4
Clonadacasey 38 B3
Clonakenny 31 F3
Clonakilty (Cloich na Coillte) 41 F3
Clonalig 22 B4
Clonalvy 29 E2
Clonard 32 A2
Clonaslee 32 A2
Clonavoe 28 C2
Clonbern 26 B3
Clonbulloge 28 C2
Clonbur (An Fhairche) 25 E2
Cloncurry 29 D3
Clondalkin 29 E4

Clondrohid 41 F2
Clondulane 42 C1
Clonea 38 A3
Clonee 29 E3
Cloneen 38 A2
Clonegal 33 D4
Clonelly 21 D1
Clones (Cluain Eois) 21 F3
Clonevin 39 E1
Cloney 32 C2
Clonfert 27 D4
Clonmany 15 E1
Clonmel (Cluain Meala) 38 A2
Clonmellon 28 C2
Clonmore Carlow 33 D4
Clonmore Offaly 28 B4
Clonmore Tipperary 31 F4
Clonmult 42 C1
Clonoe 22 B1
Clonony 27 E4
Clonoon 31 D3
Clonoulty 37 F2
Clonroche 39 D2
Clonsilla 29 E4
Clontarf 29 F4
Clontibret 22 A3
Clontygora 22 C4
Clonvaraghan 23 D2
Clonygowan 32 B2
Cloonacool 20 A4
Cloonagh 28 A1
Cloonboo (Cluain Bú) 26 A4
Cloondaff 19 E4
Cloondara 27 E2
Cloone 27 E1
Clooneagh 27 E1
Clooney 14 B4
Cloonfad Roscommon 26 B2
Cloonfad Roscommon 27 D4
Cloonfallagh 26 A1
Cloonfinish 20 A4
Cloonglasney 19 F3
Cloonkeen 19 E4
Cloonkeevy 20 B4
Cloonlara 31 D4
Cloonlogh 20 B4
Cloonlusk 37 E2
Cloonmore 20 A4
Cloonteen 37 E2
Cloontia 20 A4
Cloonyquin 27 D1
Cloran 28 C2
Clough 23 E2
Cloughduv 42 A2
Cloughjordan 31 E3
Cloughmills 16 C3
Cloughmore
  (An Chloich Mhóir) 18 C4
Cloughreagh 22 C3
Clouncagh 36 B3
Cloverhill 21 F4
Cloyfin 16 B2
Cloyne 42 C2
Cluain Bú (Cloonboo) 26 A4
Cluain Eois (Clones) 21 F3
Cluain Meala (Clonmel) 38 A2
Cluainín (Manorhamilton) 20 C3
Clynacartan 40 A2
Cnocán na Líne
  (Knocknalina) 18 C2
Coachford 42 A2
Coagh 22 B1
Coalbrook 38 A1
Coalisland 22 B1
Coan 32 B4
Cobh (An Cóbh) 42 B2
Coill an tSiáin
  (Killateeaun) 25 E1
Coillan Chollaigh
  (Bailieborough) 28 C1
Colbinstown 33 D3
Colehill 28 A2
Coleraine 16 B2
Colgagh 20 B3
Collinstown 28 B2
Collon 29 D1
Collooney 20 B3
Collorus 40 C3
Colmanstown 26 B4
Colp 29 E2
Comber 23 E1
Commeen (An Coimín) 15 D4
Commons 38 A1
Compaw 16 A2
Cong 25 E2
Conlig 23 E1
Conna 42 C1
Connemara Airport 25 E3
Connolly 30 B4
Connonagh 41 F4
Convoy 15 E3
Coogue 26 B1
Cookstown 22 B1
Coolaney 20 B3
Coolattin 33 E4
Coolbaun Kilkenny 32 B4
Coolbaun Tipperary 31 E3
Coolboy 33 E4
Coolderry Monaghan 29 D1
Coolderry Offaly 31 F3
Coole Cork 42 C1
Coole Westmeath 28 B2
Coolea (Cúil Aodha) 41 E2
Coolearagh 28 C4
Coolgrange 38 B1
Coolgreany 33 E4
Coolkeeragh 15 F2
Coollegreane 20 C3
Coolmeen 35 E2

Coolmore 20 C1
Coolock 29 F4
Coolquoy 29 E3
Coolrain 32 A3
Coolroebeg 38 B2
Cooneen 21 F2
Cooraclare 30 A4
Cootehall 20 C4
Cootehill 22 A4
Coppanrush 28 A4
Cor an Dola
  (Corrandulla) 26 A4
Cora Chaitlín
  (Newmarket-on-Fergus) 30 B4
Cora Droma Rúisc
  (Carrick-on-Shannon) 27 D1
Coralstown 28 B3
Corbally
  Newry, Mourne & Down 23 E3
Corbally Sligo 19 E3
Corcaigh (Cork) 42 B2
Corclogh (An Chorrchloch) 18 C2
Corcreeghagh 22 B4
Cordal 35 E4
Cordarragh 25 E1
Corderry 15 D3
Corduff Fingal 29 E3
Corduff Monaghan 22 A4
Corgary 15 E3
Corglass 20 C4
Cork (Corcaigh) 42 B2
Cork International Airport 42 B2
Corkey 16 B2
Corlea 27 E2
Corlee 19 F3
Corlismore 28 A1
Corlough 21 E2
Cornafannoge 21 E2
Cornafean 28 A1
Cornafulla 27 E4
Cornageeha 27 E1
Cornamaddy 22 A1
Cornamona (Corr na Móna) 25 E2
Coranagh 26 A2
Corndarragh 28 A4
Corofin Clare 30 B3
Corofin Galway 26 B3
Corr na Móna (Cornamona) 25 E2
Corraleehan 21 D4
Corrandulla (Cor an Dola) 26 A4
Corraneary 22 A4
Corraree 27 E4
Corraun (An Corrán) 18 C4
Corrawaleen 21 D4
Corries Cross 38 C1
Corriga 28 A1
Corroy 19 F3
Corry 20 C3
Cortown 28 C2
Corvally 22 A4
Costelloe (Casla) 25 D3
Courtmacsherry 42 A3
Courtown 39 E1
Cox's Hill 22 B4
Craanford 33 E4
Craig 15 F4
Craigantlet 23 E1
Craigavole 16 B3
Craigavon 22 C2
Craigdarragh 15 F3
Craigs 16 C3
Cranagh 15 F4
Cranfield 23 D4
Cranford (Creamhghort) 15 D2
Crann 22 A3
Crannogeboy 14 B4
Crannoge 22 A1
Cranny 30 A4
Crataloe 35 E3
Cratloe 30 C4
Craughwell 26 B4
Crazy Corner 28 B3
Creagh Cork 41 E4
Creagh Fermanagh & Omagh 21 E2
Creaghadoos 15 E3
Creaghanroe 22 B3
Creamhghort (Cranford) 15 D2
Crean's Cross Roads 42 A1
Crecora 36 C2
Creegh 30 A4
Creeslough 15 D2
Creevagh Mayo 19 E2
Creevagh Monaghan 22 A3
Creeves 36 B2
Cregg Cork 41 F4
Cregg Sligo 20 B3
Creggan
  Newry, Mourne & Down 22 B4
Creggan
  Fermanagh & Omagh 22 A1
Cregganbaun 25 D1
Creggaslin 26 C2
Creggs 26 C2
Crettyard 32 B4
Crilly 22 A2
Crindle 16 A2
Crinkill 31 F2
Croagh Donegal 14 B4
Croagh Limerick 36 C2
Crockbane 16 A4
Crockets Town 19 F2
Croghan Offaly 28 B4
Croghan Roscommon 27 D1
Crois Mhaoilíona
  (Crossmolina) 19 E3
Croithlí (Crolly) 14 C3
Crolly (Croithlí) 14 C3
Cromadh (Croom) 36 C2
Cromane 40 C1

Cromkill 16 C4
Crookedwood 28 B2
Crookhaven 40 C4
Crookstown Cork 42 A2
Crookstown Kildare 33 D3
Croom (Cromadh) 36 C2
Cross Clare 35 D2
Cross Mayo 25 E2
Cross Waterford 43 D1
Cross Barry 42 A2
Cross Keys Cavan 28 B1
Cross Keys Meath 29 D3
Crossabeg 39 E2
Crossakeel 28 C2
Crossconnell 27 D4
Crossdoney 21 E4
Crossea 28 A2
Crosserlough 28 B1
Crossgar 23 E2
Crossgare 16 B2
Crosshaven
  (Bun an Tábhairne) 42 B2
Crossmaglen 22 B4
Crossmolina
  (Crois Mhaoilíona) 19 E3
Crossna 20 C4
Crossnacreevy 23 D1
Crosspatrick Kilkenny 32 A4
Crosspatrick Wicklow 33 E4
Crossreagh Cavan 28 C1
Crossreagh Monaghan 21 F3
Crumlin
  Antrim & Newtownabbey 22 C1
Crumlin Dublin City 29 E4
Crusheen 30 B3
Cúil an tSúdaire
  (Portarlington) 32 B2
Cúil Aodha (Coolea) 41 E2
Cuilmore Mayo 20 A4
Cuilmore Mayo 25 D1
Culcavy 23 D1
Culdaff 15 F1
Culfadda 20 B4
Culkey 21 E2
Cullahill 32 A4
Cullane 37 D3
Cullaville 22 B4
Culleens 19 F2
Cullen Cork 41 F1
Cullen Tipperary 37 E3
Cullenstown Wexford 39 D2
Cullenstown Wexford 39 D3
Cullin 19 F3
Cullybackey 16 C3
Cullyfad 27 E2
Cullyhanna 22 B3
Culmore 15 F2
Culnady 16 B3
Culnagrew 16 B3
Cunningburn 23 E1
Curracloe 39 E2
Curragh
  Newry, Mourne & Down 23 E2
Curragh Waterford 43 D2
Curragh Camp 32 C2
Curragh West 26 C2
Curragha 29 E3
Curraghbonaun 20 A4
Curraghboy 27 D3
Curraghroe 27 D2
Curraglass 42 C1
Curraheen 42 B2
Curran 16 B4
Currans 35 D4
Curraun 25 E2
Curreeny 31 E4
Currow 35 D4
Curry Mayo 26 A2
Curry Sligo 20 A4
Curryglass 40 C3
Cushendall 17 D2
Cushendun 17 D2
Cushina 32 B2

D
Daingean 28 B4
Dalkey 29 F4
Dalystown Galway 31 D2
Dalystown Westmeath 28 B3
Damastown 29 E3
Damerstown 32 B4
Damhead 16 B2
Damhliag (Duleek) 29 E2
Danesfort 38 B1
Dangan Cork 42 C2
Dangan Kilkenny 38 B3
Darby's Bridge 40 B1
Darkley 22 B3
Darragh 30 B4
Darragh Cross 23 E2
Dartans 15 E4
Davidstown 39 D2
Delgany 33 F2
Delvin 28 C2
Dernagree 41 F1
Derreen Clare 30 A2
Derreen Mayo 19 E3
Derreendaragh 40 C2
Derreeny 40 C3
Derreeny Bridge 41 E3
Derrinturn 28 C4
Derry (Londonderry)
  Derry City & Strabane 15 F3
Derry Sligo 20 B4
Derryad 21 E3
Derrybeg (Doirí Beaga) 14 C2
Derryboy 23 E2
Derrybrien 30 C2
Derrychrier 16 A3

Derrychrin 22 B1
Derrydonnell 26 B4
Derrydruel (Doire Eadarúil) 14 B3
Derrydrumuck 22 C3
Derryerglinna
  (Doire Oirthir Ghlinne) 25 E2
Derryfadda 32 A4
Derrygolan 28 B4
Derrygonnelly 21 D2
Derrygoolin 31 D3
Derrygorry 22 A2
Derrygrogan 28 B4
Derryhale 22 B2
Derryharney 21 E3
Derryhaw 22 A2
Derryhillagh 21 E3
Derrykeevan 22 B2
Derrykeighan 16 B2
Derrylane 21 E4
Derrylea 24 C2
Derryleckagh 22 C3
Derrylin 21 E3
Derrymacash 22 C1
Derrymore Armagh City,
  Banbridge & Craigavon 22 C1
Derrymore Kerry 34 C4
Derrynacaheragh 41 E3
Derrynacreeve 21 D3
Derryneen (Doire Fhinín) 25 D2
Derrynoose 22 B3
Derryrush (Doire Iorrais) 25 D3
Derrytrasna 22 B1
Derryvane 22 B1
Derryvohy 26 A1
Derver 28 C1
Dervock 16 B2
Desertmartin 16 B4
Dingle (An Daingean) 34 B4
Doagh (Dumhach) Donegal 15 D1
Doagh
  Antrim & Newtownabbey 17 D4
Doagh Beg (Dumhaigh Bhig) 15 E1
Doire Eadarúil
  (Derrydruel) 14 B3
Doire Fhinín (Derryneen) 25 D2
Doire Iorrais (Derryrush) 25 D3
Doire Oirthir Ghlinne
  (Derryerglinna) 25 E2
Doirí Beaga (Derrybeg) 14 C2
Dolla 31 E4
Dollingstown 22 C2
Donabate 29 F3
Donacarney 29 E2
Donadea 29 D4
Donagh 21 E3
Donaghadee 23 E1
Donaghcloney 22 C2
Donaghey 22 B1
Donaghmore Mid Ulster 22 A1
Donaghmore Laois 32 A3
Donaghmore Meath 29 E3
Donaghmoyne 28 B4
Donard Wexford 39 D2
Donard Wicklow 33 D3
Donaskeagh 37 E2
Donegal (Dún na nGall) 20 C1
Donegal Airport 14 B2
Donemana 15 F3
Doneraile 36 C4
Donohill 37 E2
Donore 29 E2
Donoughmore 42 A1
Dooagh 18 C3
Doobehy 19 E2
Doocastle 20 A4
Doochary
  (An Dúchoraidh) 14 C3
Dooega (Dumha Éige) 18 C4
Doogary 21 E4
Dooghbeg
  (An Dumhach Bheag) 19 D4
Doogort 18 C3
Doohamlet 22 A3
Doohooma
  (Dumha Thuama) 18 C3
Dooish 21 E1
Doolin 30 A3
Doon Galway 26 C4
Doon Limerick 37 E2
Doona 18 C3
Doonaha 35 D2
Doonbeg 35 D1
Doonloughan 24 B2
Dooradoyle 36 C2
Dooyork (Dumha Dhearc) 18 C3
Dough 40 C4
Douglas 42 B2
Downhill 16 A2
Downies (Na Dúnaibh) 15 D2
Downpatrick 23 E2
Dowra 20 C3
Dowth 29 E2
Drangan 38 A1
Draperstown 16 B4
Dreenagh 34 C3
Dreenan 21 D1
Drimmo 32 A3
Drimoleague 41 E3
Drinagh Cork 41 F3
Drinagh Wexford 39 E3
Dring 28 A1
Dripsey 42 A2
Drishane Bridge 41 D4
Drogheda (Droichead Átha) 29 E2
Droichead Abhann
  (Sixmilebridge) 30 C4
Droichead Átha (Drogheda) 29 E2
Droichead Lios an tSonnaigh
  (Lissatinnig Bridge) 40 C2

Droichead na Bandan
(Bandon) **42** A3
Droichead Nua
(Newbridge) **33** D2
Drom *Kerry* **41** E1
Drom *Tipperary* **31** F4
Dromagh **41** F1
Dromahair **20** C3
Dromara **23** D2
Dromard **20** A3
Dromin *Limerick* **37** D3
Dromin *Louth* **29** E1
Dromina **36** C4
Dromineer **31** E3
Dromiskin **29** E1
Drommahane **42** A1
Dromod **27** E1
Dromore *Armagh City,
Banbridge & Craigavon* **23** D2
Dromore *Cork* **41** E3
Dromore
*Causeway Coast & Glens* **16** A2
Dromore
*Fermanagh & Omagh* **21** E1
Dromore West **20** A3
Drum *Mayo* **26** A1
Drum *Monaghan* **21** F4
Drum *Sligo* **20** B2
Drumacrib **22** B3
Drumadoon **16** C1
Drumahoe **15** F3
Drumalig **23** E1
Drumandoora **30** C3
Drumanespick **28** C1
Drumaness **23** D2
Drumaroad **23** D2
Drumatober **31** D2
Drumbane **37** F2
Drumbilla **22** C4
Drumbo **23** D1
Drumbologe **15** D3
Drumboy **21** D2
Drumbroagh **22** B4
Drumcar **29** E1
Drumcard **21** D3
Drumcliff **20** B2
Drumcollogher **36** B3
Drumcondra *Dublin City* **29** E4
Drumcondra *Meath* **29** D1
Drumcong **21** D4
Drumcree **28** B2
Drumduff **21** E2
Drumfad **23** F1
Drumfea **38** C1
Drumfin **20** B3
Drumfree **15** E2
Drumfries **15** E2
Drumgath **22** C3
Drumgoon **22** A4
Drumhirk **23** E1
Drumintee **22** C4
Drumkeary **31** D2
Drumkeen **37** D2
Drumkeeran **20** C3
Drumlea **21** D4
Drumlegagh **15** E4
Drumlish **27** E1
Drumlonan **21** D4
Drumlosh **27** E4
Drummacabranagher **21** E3
Drummacool **20** C3
Drummannon **22** B2
Drummenny **20** C1
Drummin **38** C2
Drummullin **27** D1
Drumnacross **15** D3
Drumnafinnila **20** C3
Drumnakilly **21** F1
Drumnalost **14** C4
Drumone **28** B2
Drumquin **21** E1
Drumraney **27** E3
Drumree **29** D3
Drumsallen **22** B2
Drumshanbo **20** C4
Drumshancorick **21** F3
Drumsna **27** D1
Drumsurn **16** A3
Drung **21** F4
Duagh **35** E3
Dually **37** F2
Dublin (Baile Átha Cliath) **29** E4
Dublin Airport **29** E4
Duleek (Damhliag) **29** E2
Dulick **30** B4
Dullaghan **21** E1
Dumha Dhearc
(Dooyork) **18** C3
Dumha Éige (Dooega) **18** C4
Dumha Thuama
(Doohooma) **18** C3
Dumhach (Doagh) **15** D1
Dumhaigh Bhig
(Doagh Beg) **15** E1
Dún an Rí (Kingscourt) **28** C1
Dún Búinne (Dunboyne) **29** E3
Dún Chaoin (Dunquin) **34** A4
Dún Dealgan (Dundalk) **22** C4
Dún Garbhán
(Dungarvan) **38** A4
Dún Géagáin
(Dungeagan) **40** B2
Dún Laoghaire **29** F4
Dún Léire (Dunleer) **29** E1
Dún Lúiche (Dunlewy) **14** C3
Dún Mánmhaí
(Dunmanway) **41** F3
Dún Mór (Dunmore East) **38** C4
Dún na nGall (Donegal) **20** C1
Dunaff **15** E1

Dunaghy **16** B2
Dunamon **26** C2
Dunany **29** E1
Dunbell **38** B1
Dunboyne (Dún Búinne) **29** E3
Duncannon **38** C3
Duncannon Bridge **41** E1
Duncormick **39** D3
Dundalk (Dún Dealgan) **22** C4
Dunderry **29** D2
Dundonald **23** E1
Dundrod **23** D1
Dundrum *Armagh City,
Banbridge & Craigavon* **22** B3
Dundrum
*Newry, Mourne & Down* **23** E3
Dundrum
*Dún Laoghaire-Rathdown* **29** E4
Dundrum *Tipperary* **37** E2
Dunfanaghy **15** D2
Dungannon **22** A1
Dungarvan *Kilkenny* **38** C1
Dungarvan (Dún Garbhán)
*Waterford* **38** A4
Dungeagan
(Dún Géagáin) **40** B2
Dungiven **16** A3
Dungloe
(An Clochán Liath) **14** B3
Dungourney **42** C2
Dunhill **38** B3
Duniry **31** D2
Dunkerrin **31** F3
Dunkineely **20** B1
Dunkitt **38** B3
Dunlavin **33** D2
Dunleer (Dún Léire) **29** E1
Dunlewy (Dún Lúiche) **14** C3
Dunloy **16** C3
Dunmanus **41** D4
Dunmanway
(Dún Mánmhaí) **41** F3
Dunmoon **43** D1
Dunmore
*Newry, Mourne & Down* **23** D2
Dunmore *Galway* **26** C1
Dunmore East (Dún Mór) **38** C4
Dunmurry **23** D1
Dunnamaggan **38** B2
Dunnamore **16** A4
Dunquin (Dún Chaoin) **34** A4
Dunsany **29** D3
Dunshaughlin **29** D3
Durlas (Thurles) **37** F2
Durrow *Laois* **32** B4
Durrow *Offaly* **28** A4
Durrus **41** D3
Dyan **22** A2
Dysart **28** A3
Dysert **27** D3

**E**

Éadan Doire (Edenderry) **28** C4
Eanach Dhúin
(Annaghdown) **26** A4
Earra Thíre na Binne
(Arryheernabin) **15** E1
Easky **19** F2
Eden **17** E4
Edenderry (Éadan Doire) **28** C4
Edendork **22** B1
Ederny **21** E1
Edgeworthstown **28** A2
Edmondstown **26** C1
Eglinton **15** F2
Eglish **22** A2
Eighter **28** B1
Ellistrin **15** D3
Elphin **27** D1
Elton **37** D3
Emly **37** D3
Emmoo **27** D2
Emo **32** B2
Emyvale **22** A2
Enfield **28** C3
Ennis (Inis) **30** B4
Enniscorthy (Inis Córthaidh) **39** D2
Enniscrone **19** F2
Enniskean **41** F3
Enniskerry
(Áth an Sceire) **33** F2
Enniskillen **21** E2
Ennistymon (Inis Diomáin) **30** A3
Ennybegs **27** E1
Eochaill (Youghal) **43** D2
Eoghanacht (Onaght) **25** D4
Erra **27** E2
Errill **32** A4
Ervey Cross Roads **15** F3
Eshnadarragh **21** F3
Eshnadeelada **21** F3
Esker **27** E1
Eskragh **21** F2
Eyeries **40** C3
Eyrecourt **31** E2

**F**

Fahafeelagh **38** A3
Fahamore **34** C4
Fahan **15** E2
Fahy **31** E2
Fairymount **26** C1
Falcarragh (An Fál Carrach) **14** C2
Fallmore (An Fál Mór) **18** C3
Fanore **30** A2
Fardrum **27** E4
Farnaght **27** E1
Farnanes **42** A2

Farnoge **38** C2
Farrahy **37** D4
Farran **42** A3
Farrancassidy **20** C2
Farranfore **35** D4
Feagarrid **37** F4
Feakle **30** C3
Fedamore **36** C2
Feeard **34** C2
Feenagh **36** B3
Feeny **16** A3
Feevagh **27** D3
Feltrim **29** F3
Fenagh **21** D4
Fenit **34** C4
Fennagh **32** C4
Fennor **38** B3
Feohanagh
(An Fheothanach) **34** A4
Feoonanagh **36** B3
Ferbane (An Féar Bán) **27** E4
Fermoy
(Mainistir Fhear Maí) **42** C1
Ferns **39** E1
Ferraghfad **27** E2
Ferrybank **33** F4
Ferrycarrig **39** E2
Fethard (Fiodh Ard)
*Tipperary* **38** A2
Fethard *Wexford* **38** C3
Fews **38** A3
Feystown **17** D3
Fiddown **38** B3
Fieres **35** D4
Finglas **39** E4
Finnea **28** B1
Finnis **23** D2
Finny (Fionnaithe) **25** E2
Fintona **21** F1
Fintown (Baile na Finne) **14** C3
Finuge **35** D3
Finvarra **30** B2
Finvoy **16** B3
Fiodh Ard (Fethard) **38** A2
Fionnaithe (Finny) **25** E2
Fisherhill **19** F3
Fiveslley **31** F2
Fivemilebridge **42** B2
Fivemiletown **21** F2
Flagmount **30** C3
Fontstown **32** C2
Fordstown **28** C2
Fore **28** B2
Foreglen **16** A3
Forgney **27** E3
Forkhill **22** C4
Formoyle **30** A2
Forthill **27** E3
Foulkesmill **39** D3
Fountain Cross **30** B4
Four Mile House **27** D2
Four Roads **27** D3
Foxford (Béal Easa) **19** F3
Foygh **27** E2
Foynes **36** B2
Freemount **36** B4
Frenchpark **26** C1
Freshford **32** A4
Freynestown **33** D3
Friarstown **29** E4
Frosses **20** C1
Fuerty **27** D2
Furbogh (Na Forbacha) **26** A4
Furraleigh **38** A3
Fybagh **34** C4

**G**

Gaillimh (Galway) **26** A4
Gainestown **28** B3
Gairfean (Garrivinnagh) **25** D3
Galbally *Mid Ulster* **22** A1
Galbally *Limerick* **37** D3
Galbally *Wexford* **39** D3
Galgorm **16** C3
Galmoy **32** A4
Galros **31** F2
Galway (Gaillimh) **26** A4
Gamblestown **22** C2
Gaoth Dobhair
(Gweedore) **14** C2
Gaoth Sáile (Gweesalia) **18** C3
Garadice **29** D3
Gargrim **21** D1
Garinish **40** B3
Garlowcross **29** D2
Garr **28** B4
Garrafrauns **26** B2
Garranlahan **26** B2
Garrison **20** C2
Garristown **29** E3
Garryclonagh **19** E3
Garrycullen **39** D3
Garrynageragh **38** A4
Garryvoe **43** D2
Garvagh
*Causeway Coast & Glens* **16** B3
Garvagh *Leitrim* **21** D4
Garvaghy **21** F1
Garvary **21** E2
Gattabaun **32** A4
Gaybrook **28** B3
Geashill **28** B4
Geevagh **20** C4
George Best
Belfast City Airport **23** D1
Gerahies **41** D3
Giantsgrave **37** F3
Gibstown (Baile Ghib) **29** D2

Gilford **22** C2
Gillygooly **21** F1
Glandore **41** F4
Glanmire **42** B2
Glantane **42** A1
Glanworth **37** D4
Glarryford **16** C3
Glashygolgan **15** F4
Glaslough **22** A2
Glassan **27** E3
Gleann Bhairr (Glenvar) **15** E2
Gleann Cholm Cille
(Glencolumbkille) **14** A4
Gleann Doimhin
(Glendowan) **15** D3
Gleann na Muaidhe
(Glenamoy) **19** D2
Glebe *Donegal* **15** E2
Glebe *Derry City & Strabane* **15** E4
Glen *Donegal* **15** D2
Glen *Fermanagh & Omagh* **20** C3
Glenade **20** C2
Glenahulla **37** D4
Glenamaddy **26** C2
Glenamoy
(Gleann na Muaidhe) **19** D2
Glenanne **22** B3
Glenariff **17** D2
Glenarm **17** D3
Glenavy **22** C1
Glenbeigh **40** C1
Glenboy **20** C3
Glenbrien **39** E2
Glenbrohane **37** D3
Glenbrook **42** B2
Glencar **20** B2
Glencolumbkille
(Gleann Cholm Cille) **14** A4
Glencorrib **26** A3
Glencullen **29** E4
Glenderry **34** C3
Glendowan
(Gleann Doimhin) **15** D3
Glendree **30** C3
Glenduff **37** E4
Glenealy **33** F3
Gleneely *Donegal* **15** E4
Gleneely *Donegal* **15** F1
Glenfarne **21** D3
Glenflesk **41** E1
Glengarriff **41** D3
Glengavlen **21** D3
Glengormley **17** D4
Glenhead **16** A3
Glenhull **16** A4
Gleniff **20** B2
Glenmaquin **15** E3
Glenmore *Clare* **30** A4
Glenmore *Kilkenny* **38** C2
Glenoe **17** D4
Glenshesk Bridge **16** C2
Glenstal Abbey **37** D2
Glenties (Na Gleannta) **14** C4
Glentogher **15** F1
Glenvar (Gleann Bhairr) **15** E2
Glenville **42** B1
Glin **35** E2
Glinsce (Glinsk) **24** C3
Glinsk *Donegal* **15** D2
Glinsk *Galway* **26** C2
Glinsk (Glinsce) *Galway* **24** C3
Glounthaune **42** B2
Glynn *Carlow* **38** C2
Glynn *Mid & East Antrim* **17** E4
Gneevgullia **41** E1
Gob an Choire (Achill Sound) **18** C4
Golden **37** F3
Goleen **41** D4
Goold's Cross **37** F2
Goresbridge **38** C1
Gorey (Guaire) **39** E1
Gormanston **29** E2
Gort (An Gort) **30** C2
Gort an Chalaidh
(Gortachalla) **26** A4
Gort an Choirce
(Gortahork) **14** C2
Gortachalla
(Gort an Chalaidh) **26** A4
Gortaclare **21** F1
Gortaganny **26** B1
Gortahork
(Gort an Choirce) **14** C2
Gortaleam **26** B2
Gortarevan **31** E2
Gortaroo **43** D2
Gortatlea **35** D4
Gorteen *Cavan* **28** A1
Gorteen *Galway* **26** C4
Gorteen *Galway* **26** B2
Gorteen *Kilkenny* **32** B4
Gorteen *Offaly* **32** A2
Gorteen (An Goirtín)
*Waterford* **38** A4
Gorteeny **31** D3
Gortgarrigan **20** C3
Gortin **15** F4
Gortmore (An Gort Mór) **25** D3
Gortmullan **21** E3
Gortnadeeve **26** C2
Gortnahey **16** A3
Gortnahoo **38** A1
Gortnasillagh **26** C1
Gortreagh **22** A1
Gortymadden **31** D2
Gowlaun **24** C1
Gowran **38** C1
Gracehill **16** C3
Graffy (An Ghrafaidh) **14** C4

Gráig na Manach
(Graiguenamanagh) **38** C1
Graigue **32** C4
Graiguealug **32** C4
Graiguenamanagh
(Gráig na Manach) **38** C1
Granabeg **33** E2
Granagh **36** C3
Granard **28** A1
Graney **33** D3
Grange *Kildare* **28** C4
Grange *Kilkenny* **38** B1
Grange *Limerick* **37** D2
Grange *Louth* **22** C4
Grange *Sligo* **20** B2
Grange *Tipperary* **38** A1
Grange *Tipperary* **37** F3
Grange *Waterford* **43** D1
Grange Blundel **22** B2
Grange Con **33** D3
Grange Corner **16** C3
Grangebellow **29** E1
Grangeford **33** D4
Grannagh **30** C2
Granny **38** B3
Gransha **23** D2
Granville **22** A1
Greagh **21** D4
Greenagh **42** A1
Greenan **33** E3
Greenanstown **29** E2
Greencastle *Donegal* **16** A1
Greencastle
*Newry, Mourne & Down* **23** D4
Greencastle
*Fermanagh & Omagh* **15** F4
Greenfield **25** E2
Greenisland **17** D4
Greenore **23** D4
Grevine **38** B1
Greyabbey **23** E1
Greysteel **15** F4
Greystones (Na Clocha Liathe) **33** F2
Grogan **27** E4
Groomsport **17** E4
Grousehall **28** B1
Guaire (Gorey) **39** E1
Gubaveeny **21** D3
Gulladuff **16** B4
Gurteen *Leitrim* **20** C2
Gurteen *Sligo* **20** B4
Gusserane **38** C2
Gweedore (Gaoth Dobhair) **14** C2
Gweesalia (Gaoth Sáile) **18** C3
Gyleen **42** C2

**H**

Hacketstown **33** D3
Haggardstown **22** C4
Halfway **42** A2
Halltown **29** D2
Hamilton's Bawn **22** B2
Hannahstown **23** D1
Hawkinstown **29** E2
Headford **26** A3
Helen's Bay **17** E4
Herbertstown **37** D2
Highwood **20** C4
Hill of Down **28** C3
Hillhall **23** D1
Hillsborough **23** D2
Hillstreet **27** D1
Hilltown
*Newry, Mourne & Down* **23** D3
Hilltown *Wexford* **39** D3
Hoathstown **29** D1
Hollyford **37** E2
Hollyfort **33** E4
Hollymount **26** A2
Hollywood **33** D2
Holycross *Limerick* **37** D2
Holycross *Tipperary* **37** F2
Holywell **21** D3
Holywood **23** E1
Horse & Jockey **37** F2
Horseleap *Galway* **26** B3
Horseleap *Westmeath* **28** A4
Hospital **37** D3
Howth **29** F4
Hugginstown **38** B2
Hurler's Cross **30** C4

**I**

Illies **15** F2
Inagh **30** B3
Inch *Cork* **43** D1
Inch *Cork* **42** C2
Inch *Donegal* **15** E2
Inch *Kerry* **34** C4
Inch *Tipperary* **37** E2
Inch *Wexford* **33** E4
Inchigeelagh **41** F2
Indreabhán (Inveran) **25** E3
Inis (Ennis) **30** B4
Inis Córthaidh (Enniscorthy) **39** D2
Inis Diomáin (Ennistymon) **30** A3
Inisclan **15** E4
Inishcarra **42** A2
Inishkeen **22** B4
Inishrush **16** B3
Inistioge **38** C2
Innishannon **42** A3
Inver *Donegal* **20** C1
Inver (An tInbhear) *Mayo* **18** C3
Inveran (Indreabhán) **25** E3
Ireland West Airport Knock **26** B1
Irishtown **26** B2
Irvinestown **21** E2

**61**

## J

Jamestown *Laois* 32 B2
Jamestown *Leitrim* 27 D1
Jenkinstown 22 C4
Johnstown *Kildare* 29 D4
Johnstown *Kilkenny* 32 A4
Johnstown *Meath* 29 D2
Johnstown *Wicklow* 33 F4
Johnstown *Wicklow* 33 E4
Johnstown Bridge 28 C4
Johnswell 32 B4
Jonesborough 22 C4
Julianstown 29 E2

## K

Kanturk (Ceann Toirc) 36 B4
Katesbridge 23 D2
Keadew 20 C4
Keady 22 B3
Keale 35 E3
Kealkill 41 E3
Kearney 23 F2
Keel 18 C3
Keeloges 26 C2
Keenagh *Longford* 27 E2
Keenagh *Mayo* 19 E3
Keenaghan 20 C2
Keeran 21 E1
Kells *Mid & East Antrim* 16 C4
Kells *Kerry* 40 B1
Kells *Kilkenny* 38 B1
Kells (Ceanannus Mór)
    *Meath* 28 C2
Kellysgrove 27 D4
Kenmare (Neidín) 41 D2
Kentstown 29 D2
Kerry Airport 35 D2
Kesh *Fermanagh & Omagh* 21 D1
Kesh *Sligo* 20 B4
Keshcarrigan 21 D4
Kilbaha 34 C2
Kilbane 31 D4
Kilbarry *Cork* 41 F2
Kilbarry *Waterford* 38 B3
Kilbeacanty 30 C2
Kilbeggan 28 A4
Kilbeheny 37 E4
Kilbennan 26 B3
Kilberry *Kildare* 32 C3
Kilberry *Meath* 29 D2
Kilbreedy 36 C2
Kilbricken 32 A3
Kilbride *Meath* 29 E3
Kilbride (Cill Bhríde) *Meath* 28 C2
Kilbride *Wicklow* 33 E2
Kilbride *Wicklow* 33 F3
Kilbrien 38 A3
Kilbrin 36 C4
Kilbrittain 42 A3
Kilcaimin 26 A4
Kilcar (Cill Charthaigh) 20 B1
Kilcarn 29 D2
Kilcavan 32 B2
Kilchreest 30 C2
Kilclaran 30 C3
Kilclonfert 28 B4
Kilclooney 14 B4
Kilcock (Cill Choca) 29 D4
Kilcogy 28 A1
Kilcolgan 30 C2
Kilcolman *Cork* 42 A3
Kilcolman *Limerick* 36 B2
Kilcomin 31 F3
Kilcommon *Tipperary* 31 E4
Kilcommon *Tipperary* 37 F3
Kilconly 26 C3
Kilconnel 30 A3
Kilconnell 26 C4
Kilcoo 23 D3
Kilcoole (Cill Chomhghaill) 33 F2
Kilcormac (Cill Chormaic) 31 F2
Kilcorney 41 F1
Kilcotty 39 E2
Kilcredan 43 D2
Kilcrohane 41 D4
Kilcullen (Cill Chuillin) 33 D2
Kilcully 42 B2
Kilcummin *Kerry* 41 D1
Kilcummin (Cill Chuimín)
    *Kerry* 34 B4
Kilcummin *Mayo* 19 F2
Kilcurly 22 C4
Kilcurry 22 C4
Kildalkey 28 C3
Kildangan 32 C2
Kildare (Cill Dara) 32 C2
Kildavin 39 D1
Kilderry 38 B1
Kildimo New 36 C2
Kildimo Old 36 C2
Kildorrery 37 D4
Kildress 22 A1
Kildrum 16 C4
Kilfeakle 37 E3
Kilfearagh 35 D2
Kilfenora 30 A3
Kilfinnane 37 D3
Kilfinny 36 C2
Kilflyn 35 D3
Kilgarvan 41 E2
Kilglass *Galway* 26 C3
Kilglass *Roscommon* 27 D1
Kilglass *Sligo* 19 F2
Kilgobnet *Kerry* 41 D1
Kilgobnet *Waterford* 38 A4
Kilgowan 33 D2
Kilkea 32 C3
Kilkeary 31 E4

Kilkee (Cill Chaoi) 35 D2
Kilkeel 23 D4
Kilkelly 26 B1
Kilkenny (Cill Chainnigh) 38 B1
Kilkenny West 27 E3
Kilkerrin 26 C3
Kilkieran (Cill Chiaráin) 25 D3
Kilkinamurry 23 D2
Kilkinlea 35 E3
Kilkishen 30 C4
Kilkit 22 A4
Kill *Cavan* 21 F4
Kill (An Chill) *Kildare* 29 D4
Kill *Waterford* 38 B3
Killabunane 41 D2
Killaclug West 37 D4
Killacolla 36 C3
Killadeas 21 E2
Killadoon 24 C1
Killadysert 36 B2
Killagan Bridge 16 C2
Killahy 38 B2
Killala 19 F2
Killaloe (Cill Dalua) 31 D4
Killaloo 15 F3
Killamerry 38 A2
Killann 39 D1
Killard 35 D1
Killare 28 A3
Killarga 20 C3
Killarney (Cill Airne) 41 D1
Killarone 30 C2
Killashee 27 E2
Killaskillen 28 B3
Killasser 19 F3
Killateeaun
    (Coill an tSiáin) 25 E1
Killavally 25 E1
Killavil 20 B4
Killavullen 42 B1
Killea 31 F4
Killead 16 C4
Killeagh 43 D2
Killeany (Cill Éinne) 25 D4
Killedmond 38 C1
Killeedy 36 B3
Killeen *Mid Ulster* 22 B1
Killeen
    *Newry, Mourne & Down* 22 C3
Killeenadeema 31 D2
Killeeneenmore 28 C4
Killeenleagh *Cork* 41 E3
Killeenleagh (An Chillín Liath)
    *Kerry* 40 B2
Killeevan 21 F3
Killegar 21 E4
Killeglan 27 D3
Killeigh 32 A2
Killen 15 E4
Killenagh 39 E1
Killenard 32 B2
Killenaule 38 A1
Killerrig 33 D4
Killeshandra 21 E4
Killeshill 28 B4
Killeshin 32 C4
Killeter 15 E4
Killimer 35 E2
Killimor 31 E2
Killinagh 28 C4
Killinaspick 38 B2
Killinchy 23 E1
Killinick 39 E3
Killinierin 33 E4
Killinkere 28 C1
Killinny 30 B2
Killinthomas 28 C4
Killiskey 33 F2
Killonecaha
    (Cill Ón Chatha) 40 A2
Killoran 26 C4
Killorglin (Cill Orglan) 40 C1
Killoscobe 26 C3
Killoscully 31 D4
Killough
    *Newry, Mourne & Down* 23 E3
Killough *Wicklow* 33 F2
Killowen 22 C4
Killshane 29 E3
Killucan 28 B3
Killukin 27 D1
Killultan 20 B1
Killumney 42 A2
Killurin 39 D2
Killybegs (Na Cealla Beaga) 20 B1
Killyclogher 21 F1
Killycolpy 22 B1
Killygordon 15 E4
Killylea 22 A2
Killyleagh 23 E2
Killyon 31 F2
Killyrammer 16 B2
Kilmacanogue 33 F2
Kilmacoo 33 F3
Kilmacow 38 B3
Kilmacrenan 15 D2
Kilmacteige 19 F3
Kilmacthomas 38 A3
Kilmactranny 20 C4
Kilmaganny 38 B2
Kilmaine 26 A1
Kilmainhamwood 28 C1
Kilmaley 30 B4
Kilmalkedar
    (Cill Maoilchéadair) 34 B4
Kilmallock
    (Cill Mocheallóg) 37 D3
Kilmanagh 38 A1
Kilmanahan 37 F4

Kilmeadan 38 B3
Kilmeague 28 C4
Kilmeedy 36 B3
Kilmeena 19 D4
Kilmessan 29 D3
Kilmichael 41 F2
Kilmihill 30 A4
Kilmood 23 E1
Kilmore *Armagh City,
    Banbridge & Craigavon* 22 B2
Kilmore *Clare* 31 D4
Kilmore
    *Newry, Mourne & Down* 23 E2
Kilmore *Mayo* 19 F3
Kilmore *Meath* 29 D3
Kilmore *Wexford* 39 D3
Kilmore Quay 39 D3
Kilmorna 35 E3
Kilmorony 32 C3
Kilmovee 26 B1
Kilmuckridge 39 E1
Kilmurry *Clare* 30 C4
Kilmurry *Cork* 41 F2
Kilmurry *Wicklow* 33 D3
Kilmurry McMahon 35 E2
Kilmyshall 39 D1
Kilnaboy 30 B3
Kilnagross 21 D4
Kilnaleck 28 B1
Kilnamanagh 39 E1
Kilnamartyra (Cill na Martra) 41 F2
Kilnamona 30 B3
Kilnock 26 B2
Kilpatrick 42 A3
Kilpedder 33 F2
Kilquiggin 33 D4
Kilraghts 16 C2
Kilrane 39 E3
Kilrea 16 B3
Kilrean 14 B4
Kilreekill 31 D2
Kilronan (Cill Rónáin) 25 D4
Kilross *Donegal* 15 D4
Kilross *Tipperary* 37 E3
Kilrossanty 38 A3
Kilrush (Cill Rois) 35 D2
Kilsallagh *Galway* 26 C2
Kilsallagh *Mayo* 19 D4
Kilsallaghan 29 E3
Kilsally 32 B1
Kilsaran 29 E1
Kilshanchoe 28 C4
Kilshanny 30 A3
Kilsheelan 38 A2
Kilskeer 28 C2
Kilskeery 21 E2
Kiltartan 30 C2
Kiltealy 39 D1
Kilteely 37 D2
Kiltegan 33 D3
Kilternan 29 F4
Kiltiernan 30 C2
Kiltimagh 26 A1
Kiltoom 27 D3
Kiltormer 27 D4
Kiltullagh 26 B4
Kiltyclogher 20 C2
Kilwaughter 17 D3
Kilworth 37 E4
Kilworth Camp 37 E4
Kinallen 23 D2
Kinard 35 E2
Kinawley 21 E3
Kincaslough (Cionn Caslach) 14 B2
Kincon 19 E2
Kingarrow (An Cionn Garbh) 14 C3
Kingscourt (Dún an Rí) 28 C1
Kinlough 20 C2
Kinnadoohy 24 C1
Kinnegad 28 B3
Kinnitty 31 F1
Kinsale (Cionn tSáile) 42 B3
Kinsalebeg 43 D2
Kinsaley 29 F3
Kinvara (Cinn Mhara) 25 D3
Kircubbin 23 E1
Kirkinriola 16 C3
Kirkistown 23 F2
Kishkeam 36 B4
Knappagh 25 D1
Knights Town 40 B2
Knock *Clare* 35 E2
Knock *Mayo* 26 A1
Knock *Tipperary* 32 A3
Knockaderry 36 B3
Knockainy 37 D3
Knockalough 30 A4
Knockananna 33 D3
Knockanarrigan 33 D3
Knockanevin 37 D3
Knockanore 43 D1
Knockanure 35 E3
Knockatallan 28 B1
Knockatemple 28 B1
Knockaunalour 42 B1
Knockaunnaglashy 40 C1
Knockavilla 37 E2
Knockbeg 20 B3
Knockboy 38 C3
Knockbrack 15 E3
Knockbrandon 33 E4
Knockbridge 22 B4
Knockbrit 37 F2
Knockcloghrim 16 B4
Knockcroghery 27 D3
Knockduff 15 D2
Knockerry 35 E2
Knockferry 26 A3

Knocklofty 37 F3
Knocklong 37 D3
Knockmore 19 F3
Knockmoyle *Galway* 31 D2
Knockmoyle
    *Fermanagh & Omagh* 21 F1
Knocknaboul 35 E4
Knocknacarry 17 D2
Knocknacree 32 C3
Knocknagashel 35 E3
Knocknagree 41 E1
Knocknahur 20 B3
Knocknalina
    (Cnocán na Líne) 18 C2
Knocknashangan 20 C2
Knockraha 42 B2
Knockroe 26 A3
Knocks *Fermanagh & Omagh* 21 E3
Knocks *Laois* 32 A2
Knocktopher 38 B2
Knockvicar 20 C4
Knuttery 42 B1

## L

Laban 30 C2
Labasheeda 35 E2
Lack 21 E1
Lackagh *Kildare* 32 C2
Lackagh
    *Causeway Coast & Glens* 16 A3
Lackamore 31 D4
Lackan *Carlow* 32 C4
Lackan *Roscommon* 27 D2
Lackan *Wicklow* 33 D2
Lady's Island 39 E3
Ladysbridge 42 C2
Lag 15 F1
Lagavara 25 C1
Lagganstown 37 F3
Laghey 20 C1
Laghtgeorge (Leacht Seoirse) 26 A4
Laghy Corner 22 B1
Lahardaun 19 E3
Lahinch 30 A3
Lakyle 35 E2
Lanesborough
    (Béal Átha Liag) 27 E2
Laracor 29 D3
Laragh *Kildare* 29 D4
Laragh *Monaghan* 22 A4
Laragh *Wicklow* 33 E3
Largan 19 F3
Largydonnell 20 B2
Larne 17 D3
Latnamard 22 A3
Lattin 37 E3
Latton 22 A4
Lauragh 40 C3
Laurelvale 22 C2
Laurencetown 27 D4
Lavagh 20 A4
Lavey 21 F4
Lawrencetown 22 C2
Laytown 29 E2
Leabgarrow
    (An Leadhb Gharbh) 14 B3
Leacht Seoirse (Laghtgeorge) 26 A4
Leamhcán (Lucan) 29 E4
Leamlara 42 C2
Leansaghane 35 D2
Leap 41 F4
Lecarrow 27 D3
Leckanarainey 20 C2
Leckanvy 19 D4
Leckaun 20 C3
Leckemy 15 F1
Leenane 25 D1
Leggagh 28 A1
Leggs 21 D1
Leifear (Lifford) 15 E4
Leighlinbridge 32 C4
Léim an Bhradáin
    (Leixlip) 29 E4
Leiter 28 C1
Leitir Ceanainn
    (Letterkenny) 15 D3
Leitir Mealláin
    (Lettermullen) 25 D3
Leitir Mhic an Bhaird
    (Lettermacaward) 14 B3
Leitir Móir (Lettermore) 25 D3
Leitrim *Armagh City,
    Banbridge & Craigavon* 23 D3
Leitrim *Leitrim* 20 C4
Leitrim *Longford* 27 E1
Leixlip (Léim an Bhradáin) 29 E4
Lemanaghan 27 F4
Lemybrien 38 A3
Lenamore 28 A2
Lerrig 35 D3
Leshemstown 29 D3
Letter 21 D1
Letterbarra 14 C4
Letterbreen 21 D2
Letterfrack 24 C2
Letterkeen 21 D3
Letterkenny
    (Leitir Ceanainn) 15 D3
Lettermacaward
    (Leitir Mhic an Bhaird) 14 B3
Lettermore (Leitir Móir) 25 D3
Lettermullen
    (Leitir Mealláin) 25 D3
Lettershendony 15 F3
Levally 26 B3
Licketstown 38 B3
Lifford (Leifear) 15 E4
Ligfordrum 15 F4

Ligoniel 23 D1
Limavady 16 A2
Limerick (Luimneach) 36 C2
Limerick Junction 37 E3
Lios Mór (Lismore) 43 D1
Lios Póil (Lispole) 34 B4
Lios Tuathail (Listowel) 35 D3
Lisacul 23 E1
Lisbane 23 E1
Lisbellaw 21 E3
Lisboduff 21 F4
Lisburn 23 D1
Liscannor 30 A3
Liscarney 25 D1
Liscarroll 36 C4
Liscloon 15 F3
Liscolman 16 B2
Liscooly 15 E4
Lisdoart 22 A2
Lisdoonvarna 30 A3
Lisdowney 32 B4
Lisduff *Cavan* 28 C1
Lisduff *Leitrim* 21 D4
Lisduff *Offaly* 31 F2
Lisgarode 31 D2
Lisgrea 28 B1
Lislea *Mid Ulster* 16 B3
Lislea *Newry, Mourne & Down* 22 C3
Lislevane 42 A4
Lismacaffry 28 A2
Lismaine 32 B4
Lismakin 31 F3
Lismoghry 15 E3
Lismore (Lios Mór) 43 D1
Lismoyle 27 D3
Lisnagry 37 D2
Lisnagunogue 16 B1
Lisnamuck 16 B4
Lisnarrick 21 D2
Lisnaskea 21 E3
Lispatrick 42 B3
Lispole (Lios Póil) 34 B4
Lisroe 30 B4
Lisronagh 38 A2
Lisryan 28 A2
Lissalway 26 C2
Lissatava 26 A2
Lissatinnig Bridge (Droichead
    Lios an tSonnaigh) 40 C2
Lisselton 35 D3
Lissinagroagh 20 C3
Lissiniska 20 C2
Lissycasey 30 B4
Listerlin 38 B2
Listooder 23 E2
Listowel (Lios Tuathail) 35 D3
Listry 41 D1
Lisvarrinane 37 E3
Littleton 37 F2
Lixnaw 35 D3
Loanends 17 D4
Lobinstown 29 D1
Loch an Iúir (Loughanure) 14 C3
Loch Garman (Wexford) 39 E2
Loghill 35 E3
Lohart 41 D2
Loiscreán (Loskeran) 38 A4
Lombardstown 42 A1
Londonderry (Derry) 15 F3
Longford (An Longfort)
    *Longford* 27 E2
Longford *Offaly* 31 F2
Longwood 28 C3
Lorrha 31 E2
Loskeran (Loiscreán) 38 A4
Lough Eglish 22 A4
Lough Gowna 28 A1
Lough Morne 22 A4
Loughanavally 28 A3
Loughanure (Loch an Iúir) 14 C3
Loughbrickland 22 C2
Loughduff 28 A1
Lougher 34 C4
Loughgall 22 B2
Loughglinn 26 C1
Loughguile 16 C2
Loughinisland 23 E2
Loughmacrory 21 F1
Loughmoe 31 F4
Loughrea (Baile Loch Riach) 31 D2
Loughshinny 29 F3
Louisburgh 19 D4
Louth 22 B4
Lower Ballinderry 22 C1
Lowertown *Cork* 41 D4
Lowertown
    *Derry City & Strabane* 15 F3
Lowtown 23 D2
Lucan (Leamhcán) 29 E4
Luimneach (Limerick) 36 C2
Lukeswell 38 B2
Lullymore 28 C4
Lurgan *Armagh City,
    Banbridge & Craigavon* 22 C2
Lurgan *Roscommon* 26 C1
Lurganare 22 C3
Lurganboy *Donegal* 15 E2
Lurganboy *Leitrim* 20 C2
Lurgill 22 C1
Lusca (Lusk) 29 F3
Lusk (Lusca) 29 F3
Lyracrumpane 35 D3
Lyre *Cork* 41 F3
Lyre *Cork* 42 A1
Lyrenaglogh 37 E4

**62**

## M

Maam Cross (An Teach Dóite) 25 D2
Maas 14 B4
McGregor's Corner 16 C3
Machaire Dromann
 (Magheradrumman) 15 E1
Mackan 21 E3
Macosquin 16 B2
Macroom (Maigh Chromtha) 41 F2
Maddan 22 B3
Maddockstown 38 B1
Maganey 32 C3
Maghaberry 22 C1
Maghanlawaun 40 C1
Maghera (An Machaire)
 Donegal 14 B4
Maghera
 Newry, Mourne & Down 23 D3
Maghera Mid Ulster 16 B3
Magherabane 15 E1
Magheracloone 29 D1
Magheradrumman
 (Machaire Dromann) 15 E1
Magherafelt 16 B4
Magheralin 22 C2
Magheramason 15 E1
Magheramayo 23 D3
Magheramorne 17 E4
Magheraveely 21 F3
Maghery Armagh City,
 Banbridge & Craigavon 22 B1
Maghery (An Machaire)
 Donegal 14 B3
Maghery
 Newry, Mourne & Down 23 D4
Magilligan 16 A2
Magonragh 21 E2
Maguiresbridge 21 E3
Mahoonagh 36 B3
Maigh Chromtha (Macroom) 41 F2
Maigh Cuilinn (Moycullen) 26 A4
Maigh Naud (Maynooth) 29 D4
Maiglean Rátha
 (Mountrath) 32 A3
Mainham 29 D4
Mainistir Eimhín
 (Monasterevin) 32 C2
Mainistir Fhear Maí
 (Fermoy) 42 C1
Mainistir Laoise
 (Abbeyleix) 32 B3
Mainistir na Búille (Boyle) 20 C4
Mainistir na Corann
 (Midleton) 42 C2
Mainistir na Feile
 (Abbeyfeale) 35 E3
Maíros (Moyrus) 24 C3
Máistir Gaoithe
 (Mastergeehy) 40 B2
Mala (Mallow) 42 A1
Málainn Bhig (Malin Beg) 14 A4
Málainn Mhóir (Malin More) 14 A4
Malin 15 F1
Malin Beg (Málainn Bhig) 14 A4
Malin More (Málainn Mhóir) 14 A4
Mallow (Mala) 42 A1
Mallusk 17 D4
Manorcunningham 15 E3
Manorhamilton (Cluainín) 20 C3
Manselltown 31 F4
Mansfieldstown 29 E1
Mantua 27 D1
Manulla 26 A1
Mardyke 38 A1
Markethill 22 B3
Marshalstown 39 D1
Martinstown
 Mid & East Antrim 16 C3
Martinstown Limerick 37 D3
Martinstown Meath 28 C2
Martry 28 C2
Masshill 20 A4
Mastergeehy
 (Máistir Gaoithe) 40 B2
Matehy 42 A2
Maudabawn 22 A4
Maum (An Mám) 25 D2
Mauricemills 30 B3
Mayglass 39 E3
Maynooth (Maigh Naud) 29 D4
Mayo 26 A2
Mayobridge 22 C3
May's Corner 22 C3
Maytown 22 C3
Mazetown 23 D1
Meanus 36 C2
Meelick 31 E2
Meelin 36 B4
Meenaclady
 (Mín an Chladaigh) 14 C2
Meenacross
 (Mín na Croise) 14 B3
Meenanarwa
 (Mín an Arbhair) 14 C4
Meenaneary (Mín an Aoire) 14 B4
Meenglass 15 D4
Meenlaragh (Mín Larach) 14 C2
Meenreagh 15 D4
Meentullynagarn 14 B4
Meigh 22 C3
Menlough Galway 26 C3
Menlough (Mionlach) Galway 26 A4
Middletown 22 A3
Midfield 26 A1
Midleton
 (Mainistir na Corann) 42 C2
Mile House 39 D1
Milemill 33 D2

## Milestone 37 E2 (second column)

Milestone 37 E2
Milford Carlow 32 C4
Milford Cork 36 C3
Milford Donegal 15 D2
Mill Town 16 C4
Millbay 17 E4
Millbrook Mid & East Antrim 17 D3
Millbrook Meath 28 B2
Millford 22 B2
Millisle 23 E1
Millroad 39 D3
Millstreet (Sráid an Mhuilinn)
 Cork 41 F1
Millstreet Waterford 37 F4
Milltown
 Causeway Coast & Glens 16 B2
Milltown Armagh City,
 Banbridge & Craigavon 22 C3
Milltown Cavan 21 E4
Milltown Armagh City,
 Banbridge & Craigavon 22 B1
Milltown Galway 26 C3
Milltown Galway 26 B2
Milltown Kerry 35 D4
Milltown (Baile an Mhuilinn)
 Kerry 34 B4
Milltown Kildare 32 C2
Milltown South Dublin 29 E4
Milltown
 Derry City & Strabane 15 E3
Milltown Westmeath 28 B2
Milltown Wexford 39 E1
Milltownpass 28 B3
Miltownmalbay 30 A4
Mín an Aoire (Meenaneary) 14 B4
Mín an Arbhair
 (Meenanarwa) 14 C4
Mín an Chladaigh
 (Meenaclady) 14 C2
Mín Larach (Meenlaragh) 14 C2
Mín na Croise (Meenacross) 14 B3
Minane Bridge 42 B3
Minerstown 23 E3
Mionlach (Menlough) 26 A4
Mitchelstown
 (Baile Mhistéale) 37 E4
Moate (An Móta) 27 E4
Model Village 42 A2
Modelligo 43 D1
Modreeny 31 E3
Mogeely 42 C2
Moglass 37 F2
Mohil 32 B4
Mohill 27 E1
Móinteach Milie
 (Mountmellick) 32 B2
Moira 22 C1
Monaghan (Muineachán) 22 A3
Monamolin 39 E1
Monanclogh 16 C2
Monard 37 E3
Monaseed 39 E4
Monaster 36 C2
Monasteraden 26 C1
Monasterevin
 (Mainistir Eimhín) 32 C2
Monea 21 D2
Moneen Galway 26 A3
Moneen Roscommon 27 D2
Moneycarragh 23 D3
Moneydig 16 B3
Moneygall 31 E3
Moneyglass 16 C4
Moneygold 20 B2
Moneylahan 20 B2
Moneymore Mid Ulster 16 B4
Moneymore Louth 29 E2
Moneyneany 16 A4
Moneyreagh 23 E1
Moneyslane 23 D3
Monilea 28 B3
Monivea 26 B4
Monkstown 42 B2
Monroe 28 A3
Monteith 22 D2
Montpelier 31 D4
Mooncoin 38 B3
Moone 32 C3
Moorfields 16 C4
Moortown 22 B1
Morenane 36 C3
Morley's Bridge 41 E2
Mornington 29 E2
Morpeth Bridge 37 E2
Mosney 29 E2
Mossley 17 D4
Moss-side 16 C2
Mothel 38 A3
Mount Hamilton 16 A4
Mount Norris 22 B3
Mount Nugent 28 B1
Mount Talbot 27 D3
Mount Temple 27 E3
Mount Uniacke 42 C1
Mountallen 20 C4
Mountbellew 26 C3
Mountbolus 32 A2
Mountcharles 20 C1
Mountcollins 35 E4
Mountfield 21 F1
Mountjoy Mid Ulster 22 B1
Mountjoy
 Fermanagh & Omagh 21 F1
Mountmellick
 (Móinteach Milie) 32 B2
Mountrath
 (Maiglean Rátha) 32 A3
Mountshannon 31 D3
Moville (Bun an Phobail) 16 A2
Mowhan 22 B3

## Moy 22 B2 (third column)

Moy 22 B2
Moyallon 22 C2
Moyard 24 C2
Moyarget 16 C2
Moyasta 35 D2
Moybane 21 D2
Moycullen (Maigh Cuilinn) 26 A4
Moydow 27 E2
Moygashel 22 B1
Moyglass 31 D2
Moylett 28 B1
Moylough Galway 26 C3
Moylough Sligo 20 A4
Moymore 30 C4
Moynalty 28 C1
Moynalvy 29 D3
Moyne Longford 28 A1
Moyne Roscommon 26 C1
Moyne Tipperary 31 F4
Moyne Wicklow 33 E2
Moyrus (Maíros) 24 C3
Moyvalley 28 C3
Moyvane
 (Newtown Sandes) 35 E3
Moyvore 28 A3
Moyvoughly 28 A3
Mucklon 28 C4
Muckross (Mucros) Donegal 20 B1
Muckross Kerry 41 D1
Mucros (Muckross) 20 B1
Muff 15 F2
Muine Bheag
 (Bagenalstown) 32 C4
Muineachán (Monaghan) 22 A3
Mulhuddart 28 E3
Mullabohy 22 B4
Mullabrack 22 C2
Mullach Íde (Malahide) 29 F3
Mullacrew 29 D1
Mullagh Cavan 28 C1
Mullagh Clare 30 A4
Mullagh Galway 31 D2
Mullagh Mayo 25 D1
Mullagh Meath 29 D3
Mullaghbane 22 B4
Mullaghboy 17 E3
Mullaghmassa 21 F1
Mullaghmore 20 B2
Mullaghroe 20 B4
Mullan 21 D2
Mullanacross 20 C1
Mullans Town 22 A1
Mullany's Cross 20 A4
Mullartown 23 D3
Mullen 26 C1
Mullenmeehan 28 A3
Mullinahone 28 A1
Mullinavat 38 B2
Mullingar
 (An Muileann gCearr) 28 B3
Mulrany
 (An Mhala Raithní) 19 D4
Multyfarnham 28 B2
Mungret 36 C2
Murley 21 F2
Murntown 39 E3
Murreagh (An Mhuiríoch) 34 A4
Murrens 28 B2
Murroe 37 D2
Murroogh 30 A2
Myrtleville 42 B3
Myshall 33 D4

## N

Na Bruíneacha (Breenagh) 15 D3
Na Cealla Beaga (Killybegs) 20 B1
Na Clocha Liatha
 (Greystones) 33 F2
Na Dúnaibh (Downies) 15 D2
Na Forbacha (Furbogh) 26 A4
Na Gleannta (Glenties) 14 C4
Na Sceirí (Skerries) 29 F2
Naas (An Nás) 33 D2
Nad 42 A1
Narin 14 B4
Narraghmore 32 C2
Naul 29 E2
Navan (An Uaimh) 29 D2
Neale 26 A4
Nealstown 31 F3
Neidín (Kenmare) 41 D2
Nenagh (An tAonach) 31 E4
New Birmingham 38 A1
New Buildings 15 F3
New Ferry 16 B4
New Inn Cavan 28 B1
New Inn Galway 26 C4
New Inn Laois 32 B2
New Mills 15 D3
New Ross
 (Ros Mhic Thriúin) 38 C2
New Twopothouse Village 36 C4
Newbawn 39 D2
Newbliss 21 F3
Newbridge Galway 26 C3
Newbridge (Droichead Nua)
 Kildare 33 D2
Newbridge Limerick 36 B2
Newcastle
 Newry, Mourne & Down 23 D3
Newcastle Galway 26 B4
Newcastle (An Caisleán Nua)
 South Dublin 29 D4
Newcastle Tipperary 37 F4
Newcastle Wicklow 33 F2
Newcastle West
 (An Caisleán Nua) 36 B3
Newcestown 42 A3
Newchapel Tipperary 31 E3

## Newchapel Tipperary 37 F3 (fourth column)

Newchapel Tipperary 37 F3
Newgrange 29 E2
Newinn 37 F3
Newmarket (Áth Trasna)
 Cork 36 B4
Newmarket Kilkenny 38 B2
Newmarket-on-Fergus
 (Cora Chaitlín) 30 B4
Newmill 17 D4
Newmills 22 B1
Newport Mayo 19 D4
Newport Tipperary 31 D4
Newry 22 C3
Newton Cloghans 19 F3
Newtown (An Baile Nua) Cork 36 C4
Newtown Kildare 29 D4
Newtown Laois 32 C4
Newtown Limerick 37 D3
Newtown Meath 29 D1
Newtown Offaly 31 E2
Newtown Roscommon 27 D4
Newtown Tipperary 31 E3
Newtown Tipperary 31 E3
Newtown Tipperary 37 E3
Newtown Waterford 38 B3
Newtown Crommelin 16 C3
Newtown Daly 31 D2
Newtown Forbes 27 E2
Newtown Gore 21 E4
Newtown Monasterboice 29 E1
Newtown Mount Kennedy
 (Baile an Chinnéidigh) 33 F2
Newtown Sandes
 (Moyvane) 35 E3
Newtownabbey 17 D4
Newtownards 23 E1
Newtownbutler 21 F3
Newtowncashel 27 E3
Newtowncunningham 15 E3
Newtownhamilton 22 B3
Newtownlow 28 A4
Newtownshandrum 36 C3
Newtownstewart 15 F4
Ninemilehouse 38 A2
Nobber 29 D1
Nohoval 42 B3
North Ring 42 A3
Northlands 22 A4
Noughaval 30 B3
Nurney Carlow 32 C4
Nurney Kildare 32 C2
Nutt's Corner 23 D1

## O

Oaghley 35 D3
Oatfield 30 C4
O'Briensbridge 31 D4
O'Callaghansmills 30 C4
Oghil 27 D4
Ogonelloe 31 D3
Oileán Ciarraí
 (Castleisland) 35 E4
Oilgate 39 D2
Old Head 42 B3
Old Kilcullen 33 D2
Old Ross 39 D2
Old Town Donegal 15 D3
Old Town Laois 32 B3
Old Town Roscommon 27 E4
Oldcastle 28 B1
Oldcourt Wexford 38 C2
Oldcourt Wicklow 33 E2
Oldleighlin 32 C4
Oldtown 29 E3
Omagh 21 F1
Omeath 22 C4
Onaght (Eoghanacht) 25 D4
Oola 37 E2
Oran 26 C2
Órán Mór (Oranmore) 26 A4
Oranmore (Órán Mór) 26 A4
Oristown (Baile Órthaí) 28 C2
Oritor 22 A1
Oughterard (Uachtar Ard) 25 E2
Oulart 39 E1
Ovens 42 A2
Owenbeg 19 F2
Owenbristy 30 C2
Owning 38 B2
Oysterhaven 42 B3

## P

Palace 39 D2
Palatine 32 C4
Pallas 32 B3
Pallas Gream 37 D2
Pallas Green New 37 D2
Pallaskenry 36 C2
Palmerston 29 E4
Palmerstown 29 E4
Park Derry City & Strabane 15 F3
Park Mayo 26 A1
Park Wexford 39 D3
Parkacunna 37 D4
Parkgate 17 D4
Parkmore 30 B2
Parknasilla 40 C2
Parkville 38 A2
Parteen 30 C4
Partry 25 E1
Passage 27 D3
Passage East 38 C3
Passage West (An Pasáiste) 42 B2
Patrickswell 36 C2
Paulstown 38 C1
Peake 42 A2
Peterswell 30 C2
Pettigo 21 D1

## Pharis 16 C2 (fifth column)

Pharis 16 C2
Pike of Rush Hall 32 A3
Piltown 38 B2
Pipers Town 27 E1
Pluck 15 E3
Plumbridge 15 F4
Poles 21 F4
Poll a'tSómais (Pollatomish) 19 D2
Pollagh 27 E4
Pollaphuca 33 F4
Pollatomish
 (Poll a'tSómais) 19 D2
Pollremon 26 B2
Pollshask 26 C2
Pomeroy 22 A1
Pontoon 19 F3
Port 29 E1
Port an Chlóidh (Portacloy) 19 D1
Port Durlainne (Porturlin) 19 D1
Port Lach (Portlaw) 38 B3
Port Láirge (Waterford) 38 C3
Port Laoise (Portlaoise) 32 B3
Port Mearnóg
 (Portmarnock) 29 F3
Port Omna (Portumna) 31 E2
Port Reachrann (Portrane) 29 F3
Portacloy (Port an Chlóidh) 19 D1
Portadown 22 C2
Portaferry 23 E2
Portaleen 15 F1
Portarlington
 (Cúil an tSúdaire) 32 B2
Portavogie 23 F2
Portballintrae 16 B1
Portbraddan 16 C1
Portglenone 16 B3
Portland 31 E2
Portlaoise (Port Laoise) 32 B3
Portlaw (Port Lach) 38 B3
Portmagee 40 A2
Portmarnock
 (Port Mearnóg) 29 F3
Portnoo 14 B4
Portrane (Port Reachrann) 29 F3
Portroe 31 D3
Portrunny 27 D3
Portrush 16 B1
Portsalon 15 E1
Portstewart 16 B2
Portumna (Port Omna) 31 E2
Porturlin (Port Durlainne) 19 D1
Poulanargid 41 F2
Poulnagunoge 38 A2
Poulnamucky 37 F3
Power's Cross 31 D2
Powerstown 38 C1
Poyntz Pass 22 C3
Prehen 15 F3
Priesthaggard 38 C3
Priestsessagh 15 E4
Prosperous 29 D4
Puckaun 31 E3

## Q

Querrin 35 D2
Quilty 30 A4
Quin 30 C4

## R

Raffrey 23 E2
Raghly 20 A2
Rahan 28 A4
Rahara 27 D3
Raharney 28 C3
Rahavanig 35 D2
Raheen Cork 42 A2
Raheen Limerick 36 C2
Raheen Wexford 39 D2
Raheny 29 F4
Raholp 23 E2
Rakestreet 19 E3
Raloo 17 D4
Ramelton (Ráth Mealtain) 15 E2
Ranaghan 16 A4
Randalstown 16 C4
Rapemills 31 F2
Raphoe (Ráth Bhoth) 15 E3
Rashedoge 15 D3
Rath 31 F2
Ráth Bhoth (Raphoe) 15 E3
Ráth Caola (Rathkeale) 36 B2
Ráth Cúil (Rathcoole) 29 E4
Ráth Domhnaigh
 (Rathdowney) 32 A4
Ráth Droma (Rathdrum) 33 E3
Ráth Iomgháin (Rathangan) 32 C2
Ráth Mealtain (Ramelton) 15 E2
Ráth Naoi (Rathnew) 33 F2
Rathangan (Ráth Iomgháin) 32 C2
Rathaspick 28 A2
Rathbraghan 20 B3
Rathcabban 31 E2
Rathcardan 31 F4
Rathconrath 28 A3
Rathcool 41 F1
Rathcoole (Ráth Cúil) 29 E4
Rathcor 22 C4
Rathcore 28 C3
Rathcormac 42 C1
Rathcormack 20 B2
Rathdangan 33 D3
Rathdowney
 (Ráth Domhnaigh) 32 A4
Rathdrinagh 29 D2
Rathdrum (Ráth Droma) 33 E3
Rathduff 42 A1
Rathfarnham 29 E4
Rathfeigh 29 E2

Rathfriland **22** C3
Rathgormuck **38** A3
Rathkeale (Ráth Caola) **36** B2
Rathkeevin **37** F3
Rathkenny **29** D2
Rathlee **19** F2
Rathleek **29** D4
Rathmolyon **28** C3
Rathmore *Kerry* **41** E1
Rathmore *Kildare* **33** D2
Rathmoyle **32** B3
Rathmullan **15** E2
Rathmullen **20** B4
Rathnew (Ráth Naoi) **33** F3
Rathnure **39** D2
Rathoma **19** E2
Rathowen **28** A2
Rathumney **38** C3
Rathvilla **28** B4
Rathvilly **33** D3
Rathwire **28** B3
Ratoath **29** E3
Ravensdale **22** C4
Ravernet **23** D1
Ray **15** E2
Ré na gCloichín
  (Reanaclogheen) **38** A4
Ré na nDoirí (Reananeree) **41** F2
Readypenny **29** E1
Reaghstown **29** D1
Reanaclogheen
  (Ré na gCloichín) **38** A4
Reanagowan **35** D4
Reananeree (Ré na nDoirí) **41** F2
Rear Cross **37** E3
Recess (Sraith Salach) **25** D2
Redcastle **15** F2
Redcross **33** F3
Redford **33** F2
Redgate **39** E2
Redhills **21** F4
Reen **41** D2
Reens **36** B2
Reevanagh **32** B4
Relagh Beg **28** C1
Rerrin **40** C3
Rhode **28** B4
Richhill **22** B2
Ridge **32** C4
Rineen **41** E4
Ringaskiddy **42** B2
Ringboy **23** F2
Ringsend **16** B2
Ringstown **32** A3
Ringville (An Rinn) **38** A4
Rinnananny **19** F3
Rinneen **30** A3
Rinville **26** A4
Rinvyle **24** C1
River **18** C3
Riverchapel **39** E1
Riverside **23** D4
Riverstick **42** B3
Riverstown *Cork* **42** B2
Riverstown *Sligo* **20** B4
Riverstown *Tipperary* **31** F2
Roadford **30** A3
Robertstown **28** C4
Robinstown **29** D2
Rochestown *Cork* **42** B2
Rochestown *Kilkenny* **38** C3
Rochfortbridge **28** B3
Rockbrook **29** E4
Rockchapel **36** B4
Rockcorry **22** A4
Rockhill **36** C3
Rockmills **37** D4
Rodeen **37** D4
Rooaun **31** E2
Roonah Quay **18** C4
Roosky *Mayo* **20** A4
Roosky *Roscommon* **27** E1
Ros an Mhíl (Rossaveel) **25** D3
Ros Cathail (Rosscahill) **25** E3
Ros Comáin (Roscommon) **27** D2
Ros Cré (Roscrea) **31** F3
Ros Mhic Thriúin
  (New Ross) **38** C2
Ros Muc (Rosmuck) **25** D3
Rosbercon **38** C2
Roscommon (Ros Comáin) **27** D2
Roscrea (Ros Cré) **31** F3
Rosegreen **37** F3
Rosemount **28** A3
Rosenallis **32** A2
Rosmuck (Ros Muc) **25** D3
Rosnakil **15** E2
Ross **28** B1
Ross West **19** E4
Rossaveel (Ros an Mhíl) **25** D3
Rossbeg **14** B4
Rossbrin **41** D4
Rosscahill (Ros Cathail) **25** E3
Rosscarbery **41** F4
Rosscor **20** C2
Rosses Point **20** B3
Rossgeir **15** E3
Rossinver **20** C2
Rosskelton **32** A3
Rosslare **39** E3
Rosslare Harbour **39** E3
Rosslea **21** F3
Rossmore **41** F3
Rossmult **37** F2
Rossnowlagh **20** C1
Rostellan **42** C2
Rostrevor **22** C4
Rosturk **19** D4

Roughgrange **29** E2
Roundfort **26** A2
Roundstone (Cloch na Rón) **24** C2
Roundwood **33** E2
Rousky **15** F4
Roxhill **16** C4
Royaloak **32** C4
Ruan **30** B3
Rubane **23** F1
Runnaroddan **26** C1
Rush (An Ros) **29** F3
Ryefield **28** C1
Rylane **42** A1

## S

Saggart **29** E4
St. Johnstown **15** E3
St. Margaret's **29** E3
St. Mullins **38** C2
Saintfield **23** E2
Saleen **42** C2
Salisbury **32** C3
Sallahig **40** B2
Sallins **29** D4
Sallybrook **42** B2
Sallypark **31** E4
Salrock **24** C1
Salthill **26** A4
Sandholes **22** A1
Sandymount **22** C4
Santry **29** E3
Saul **29** E2
Scaddy **23** E2
Scardaun **27** D3
Scarnagh **33** F4
Scarriff **31** D3
Scartaglin **35** E4
Scarva **22** C2
Schull **41** D4
Scolboa **16** C4
Scollogstown **23** E3
Scotch Corner **22** A3
Scotch Street **22** B2
Scotch Town **15** F4
Scotshouse **21** F4
Scotstown **22** A3
Scramoge **27** D2
Screeb (Scriob) **25** D3
Screen *Meath* **29** D2
Screen *Wexford* **39** E2
Screggan **28** A4
Scribbagh **20** C2
Scriob (Screeb) **25** D3
Seaforde **33** E2
Seapatrick **22** C2
Seneirl **16** B2
Seskinore **21** F1
Shanacashel **40** C1
Shanagarry **42** C2
Shanaglish **30** C3
Shanagolden **36** B2
Shanahoe **32** A3
Shanakill **31** F4
Shanavogh **30** A4
Shanballymore **37** D4
Shankill **29** F4
Shanlaragh **41** F3
Shannon **30** C4
Shannon Harbour **31** F2
Shannon International
  Airport **30** B4
Shannonbridge **27** D4
Shanragh **32** C3
Shantonagh **22** A4
Shanvaus **20** C3
Sharavogue **31** F3
Sheeanamore **33** E3
Shercock **22** A4
Sheskin **19** D2
Shillelagh **33** D4
Shinrone **31** F3
Shoptown **17** D4
Shrigley **23** E2
Shrule **26** A3
Silverbridge **22** B4
Silvermines **31** E4
Single Street **20** C2
Sion Mills **15** E4
Six Crosses **35** D3
Six Road Ends **23** E1
Sixmilebridge
  (Droichead Abhann) **30** C4
Sixmilecross **21** F1
Skeagh **28** A3
Skehana **32** B4
Skerries (Na Sceirí) **29** F2
Skerry **32** A2
Skibbereen (An Sciobairín) **41** E4
Skreen **20** A3
Slade **38** C4
Slane *Mid & East Antrim* **17** D3
Slane *Meath* **29** D2
Slanemore **28** A3
Slatmore **21** F2
Slievemurry **26** C3
Slieverue **38** C2
Sligeach (Sligo) **20** B3
Sligo (Sligeach) **20** B3
Smarmore **29** D1
Smerwick
  (Ard na Caithne) **34** A4
Smithborough **21** F3
Smithstown **32** B4
Snave Bridge **41** D3
Sneem **40** C2
Soldierstown **22** C1
Sord (Swords) **29** E3
South Marsh **22** C4
Spa **34** C4

Spamount **15** E4
Spanish Point **30** A4
Spiddal (An Spidéal) **25** E3
Spink **32** B3
Spittaltown **28** A4
Srah (An tSraith) **25** E1
Srahmore **19** D3
Srahnamanragh Bridge **18** C3
Sráid an Mhuilinn
  (Millstreet) **41** F1
Sraith Salach (Recess) **25** D2
Srath an Bhealaigh
  (Stravally) **14** B4
Srath an Urláir (Stranorlar) **15** D4
Stabannan **29** D1
Stackallen **29** D2
Staffordstown **16** C4
Stamullen **29** E2
Staplestown **29** D4
Stepaside **29** E4
Stewartstown **22** B1
Stillorgan **29** E4
Stonepark **27** E2
Stoneyford **38** B1
Stonyford **23** D1
Strabane **15** E4
Stradbally (An Sráidbhaile)
  *Kerry* **34** B4
Stradbally (An tSraidbhaile)
  *Laois* **32** B3
Stradbally *Waterford* **38** A4
Strade **26** A1
Stradone **21** F4
Straffan **29** D4
Strahart **39** D1
Straid *Mid & East Antrim* **16** C4
Straid *Donegal* **15** E1
Straid
  *Causeway Coast & Glens* **16** C1
Straid
  *Antrim & Newtownabbey* **17** D4
Straidkilly **17** D3
Strand **36** B3
Strandhill **20** B3
Strangford **23** E2
Stranocum **16** C2
Stranoodan **22** A3
Stranorlar
  (Srath an Urláir) **15** D4
Stratford **33** D1
Stravally
  (Srath an Bhealaigh) **14** B4
Streamstown **28** A3
Street **28** A2
Strokestown **27** D1
Stroove **16** A1
Stuake **42** A1
Summer Cove **42** B3
Summerhill *Meath* **29** D1
Summerhill *Meath* **29** D3
Suncroft **32** C2
Swan **32** B3
Swanlinbar **21** D3
Swan's Cross Roads **22** A3
Swatragh **16** B3
Swinford
  (Béal Átha na Muice) **26** A1
Swords (Sord) **29** E3

## T

Tacumshane **39** E3
Taghmon **39** D3
Tagoat **39** E3
Tahilla **40** C2
Talbotstown **33** D3
Tallaght **29** E4
Tallanstown **29** D1
Tallow **42** C1
Tallowbridge **42** C1
Tamlaght **21** E2
Tamlaght O'Crilly **16** B3
Tamnamore **22** B1
Tandragee **22** C2
Tang **29** D3
Tangaveane
  (An tSeanga Mheáin) **14** C4
Tara **29** D3
Tarbert **35** E2
Tassagh **22** B3
Taughblane **23** D2
Taur **36** B4
Tawny **15** D2
Tawnyinah **26** B1
Tawnylea **20** C3
Taylor's Cross **31** F2
Tedavnet **22** A3
Teelin (Teileann) **20** A1
Teemore **21** E3
Teeranearagh
  (An Trianlarach) **40** B2
Teerelton **41** F2
Teermaclane **30** B4
Teeromoyle **40** B1
Teevurcher **28** C2
Teileann (Teelin) **20** A1
Tellarought **38** C2
Temple **23** D1
Templeboy **20** A3
Templederry **31** E4
Templeglantine **35** E3
Templemartin **42** A2
Templemore
  (An Teampall Mór) **31** F4
Templenoe **41** D2
Templepatrick **17** D4
Templeshambo **39** D1
Templetown **38** C3

Templetuohy **31** F4
Tempo **21** E2
Termon *Cavan* **28** B1
Termon *Clare* **35** D2
Termon (An Tearmann)
  *Donegal* **15** D2
Termonbarry **27** E2
Termonfeckin **29** E1
Terrydremont **16** A3
Terryglass **31** E2
Tevrin **28** B3
The Bush **22** B1
The Diamond *Mid Ulster* **22** B1
The Diamond
  *Fermanagh & Omagh* **21** F1
The Downs **28** B3
The Drones **16** C2
The Five Roads **29** E3
The Harrow **39** E1
The Leap **39** D2
The Loup **16** B4
The Pigeons **27** E3
The Pike *Tipperary* **31** E3
The Pike *Waterford* **43** D1
The Rock **22** A1
The Rower **38** C2
The Sheddings **17** D3
The Six Towns **16** A4
The Spa **23** D2
The Sweep **38** B2
Thomastown (Baile Mhic Andáin)
  *Kilkenny* **38** B1
Thomastown *Meath* **28** C1
Thomastown *Tipperary* **37** E3
Thomastown *Westmeath* **28** B3
Three Mile House **22** A3
Threecastles **32** B4
Thurles (Durlas) **37** F2
Tibohine **26** C1
Tievemore **21** D1
Tildarg **17** D4
Timahoe *Kildare* **28** C4
Timahoe *Laois* **32** B3
Timoleague **42** A3
Timolin **32** C3
Tinahely **33** E4
Tinnahinch **38** C1
Tinryland **32** C4
Tiobraid Árann (Tipperary) **37** E3
Tipperary (Tiobraid Árann) **37** E3
Tircur **15** F4
Tirnaneill **22** A3
Tirneevin **30** C3
Toames **41** F2
Tobar an Choire
  (Tobercurry) **20** A4
Tober **28** A4
Tobercurry
  (Tobar an Choire) **20** A4
Tobermore **16** B4
Toberscanavan **20** B3
Toem **37** E2
Togher *Cork* **41** F3
Togher *Louth* **29** E1
Togher *Meath* **28** C3
Togher *Offaly* **27** E4
Tóin an tSeanbhaile
  (Tonatanvally) **18** C3
Tombrack **39** D1
Tomdarragh **33** E2
Tomhaggard **39** E3
Tonatanvally
  (Tóin an tSeanbhaile) **18** C3
Toneel **21** D2
Tonyduff **22** A4
Tooban **15** E2
Toomard **26** C3
Toombeola (Tuaim Beola) **24** C2
Toome **16** B4
Toomevara **31** E4
Toor **31** E4
Tooraneena **37** F4
Toorarree **35** E2
Tooreencahill **35** E4
Toorlestraun **20** A4
Toormore **41** D4
Torr **17** D1
Tourmakeady
  (Tuar Mhic Éadaigh) **25** E1
Tournafulla **36** B3
Tower **42** A2
Trá Li (Tralee) **35** D4
Trá Mhór (Tramore) **38** B3
Tracton **42** B3
Trafrask **41** D3
Tralee (Trá Li) **35** D4
Tramore (Trá Mhór) **38** B3
Trean (An Trian) **25** E1
Treantagh **15** D3
Treehoo **21** F4
Trien **26** C1
Trillick **21** E2
Trim (Baile Átha Troim) **29** D2
Tristia (Troiste) **18** C2
Troiste (Tristia) **18** C2
Trory **21** E2
Trust **21** E2
Tuaim (Tuam) **26** B3
Tuaim Beola
  (Toombeola) **24** C2
Tuam (Tuaim) **26** B3
Tuamgraney **31** D3
Tuar Mhic Éadaigh
  (Tourmakeady) **25** E1
Tubber **30** C3
Tubbrid *Kilkenny* **32** A4
Tubbrid *Tipperary* **37** F4
Tulach Mhór (Tullamore) **28** A4
Tulla *Clare* **30** B2
Tulla *Clare* **30** C4

Tullaghan **20** B2
Tullaghanstown **28** C2
Tullaghought **38** B2
Tullaherin **38** B1
Tullamore (Tulach Mhór) **28** A4
Tullaroan **38** A1
Tullcoe **21** F4
Tullig **40** C1
Tullintrain **15** F3
Tullogher **38** C2
Tullow (An Tulch) **33** D4
Tully *Donegal* **20** C1
Tully *Fermanagh & Omagh* **21** E3
Tully *Roscommon* **27** D2
Tully Cross **24** C1
Tullyallen **22** C4
Tullycanna **39** D3
Tullycar **15** D4
Tullyconnaught **22** C2
Tullycooly **20** C3
Tullyhogue **22** B1
Tullyhonwar **14** C4
Tullylease **36** B4
Tullynascreen **20** C3
Tullyrossmearan **21** D2
Tullyvin **21** F4
Tullyvoos **20** C1
Tulrohaun **26** B2
Tulsk **27** D1
Tummery **21** E1
Turlough *Clare* **30** B2
Turlough *Mayo* **26** A1
Turloughmore **26** B4
Turreen **27** E2
Two Mile Borris **37** F2
Two Mile Bridge **32** B2
Two Mile House **33** D2
Twomilebridge **38** A2
Tyholland **22** A3
Tylas **23** D3
Tynagh **31** D2
Tynan **22** A2
Tyrella **23** E3
Tyrrellspass **28** B4

## U

Uachtar Ard (Oughterard) **25** E2
Ummer **21** E2
Unionhall **41** F4
Upper Ballinderry **22** C1
Upperchurch **31** E4
Upperlands **16** B3
Urglin or Rutland **32** C4
Urhin **40** C3
Urlaur **26** B1
Urlingford **32** A4

## V

Valleymount **33** D2
Ventry (Ceann Trá) **34** A4
Vicarstown **32** C2
Victoria Bridge **15** E4
Villierstown **43** D1
Virginia **28** C1
Vow **16** B3

## W

Waddingtown **39** D3
Walshestown **39** E3
Walshtown **42** C1
Waringsford **23** D2
Waringstown **22** C2
Warrenpoint **22** C4
Waterfall **42** B2
Waterford (Port Láirge) **38** C3
Waterford Airport **38** C3
Watergrasshill **42** B1
Waterville (An Coireán) **40** B2
Wattlebridge **21** F3
Welchtown **15** D4
Wellingtonbridge **39** D3
West Town (An Baile Thiar) **14** C1
Westport (Cathair na Mart) **19** D4
Westport Quay **19** D4
Wexford (Loch Garman) **39** E2
Wheathill **21** D3
White Hall **41** E4
Whitecastle **15** F2
Whitechurch *Cork* **42** B1
Whitechurch *Waterford* **43** D1
Whitecross **22** B3
Whitegate *Clare* **31** D3
Whitegate *Cork* **42** C2
Whitehall **32** B2
Whitehead **17** E4
Whitehill **21** E2
Whitesides Corner **16** C4
Wicklow (Cill Mhantáin) **33** F3
Wilkinstown (Baile Uilcín) **29** D2
Willbrook **30** B3
Williamstown **26** C2
Windgap **38** B2
Wolfhill **32** B2
Woodenbridge **33** E4
Woodford **31** D2
Woodlawn **26** C4
Woodsgift **32** A4
Woodstown **38** C3
Woodtown **29** D1

## Y

Yellow Furze **29** D2
Youghal (Eochaill) **43** D2